Climate Change From Pole to Pole

BIOLOGY INVESTIGATIONS

Climate Change From Pole to Pole

BIOLOGY INVESTIGATIONS

▲ Juanita Constible, Luke Sandro, and Richard E. Lee, Jr.

NSTApress

National Science Teachers Association

Arlington, Virginia

National Science Teachers Association

Claire Reinburg, Director
Jennifer Horak, Managing Editor
Judy Cusick, Senior Editor
Andrew Cocke, Associate Editor
Betty Smith, Associate Editor

ART AND DESIGN
Will Thomas, Jr., Director
Tim French, Senior Designer, cover and interior design

PRINTING AND PRODUCTION
Catherine Lorrain, Director

SCILINKS
Tyson Brown, Director
Virginie L. Chokouanga, Customer Service and Database Coordinator

NATIONAL SCIENCE TEACHERS ASSOCIATION
Francis Q. Eberle, Executive Director
David Beacom, Publisher

LIBRARY OF CONGRESS CATALOGING-IN-PUBLICATION DATA

Constible, Juanita M.
  Climate change from Pole to Pole : biology investigations / by Juanita M. Constible, Luke H. Sandro, and Richard E. Lee, Jr.
     p. cm.
  Includes bibliographical references and index.
  ISBN 978-1-933531-23-6
  1. Climatic changes--Study and teaching (Secondary) 2. Human ecology--Study and teaching (Secondary) 3. Climatic changes--Study and teaching (Secondary)--Activity programs. I. Sandro, Luke H. II. Lee, Richard E. III. Title.
  QC981.8.C5C697 2008
  577.2'20712--dc22
                        2008024338

NSTA is committed to publishing material that promotes the best in inquiry-based science education. However, conditions of actual use may vary, and the safety procedures and practices described in this book are intended to serve only as a guide. Additional precautionary measures may be required. NSTA and the authors do not warrant or represent that the procedures and practices in this book meet any safety code or standard of federal, state, or local regulations. NSTA and the authors disclaim any liability for personal injury or damage to property arising out of or relating to the use of this book, including any of the recommendations, instructions, or materials contained therein.

Featuring sciLINKS®—a new way of connecting text and the internet. Up-to-the minute online content, classroom ideas, and other materials are just a click away. For more information go to *www.scilinks.org/faq/ moreinformation.asp.*

Contents

Part I—The Science of Climate Change

Part II—Climate Change Case Studies

CHAPTER 6—POPULATION PERIL
Polar bears decline in the Canadian Arctic

CHAPTER 7—CARRION: IT'S WHAT'S FOR DINNER
Wolves reduce the impact of climate change

CHAPTER 10—CRUEL, CRUEL SUMMER
Heat waves increase from pole to pole

How to Use This Book

"One of the biggest obstacles to making a start on climate change is that it has become a cliché before it has even been understood."
—Tim Flannery, *The Weather Makers,* 2005

"In order to understand scientific issues, you have to have a clear picture of what science is and what scientists do."
—Moti Ben-Ari, *Just a Theory: Exploring the Nature of Science,* 2005

This book is about the *science* of climate change, not the political issues it has generated. Part of our goal in writing it is to help you and your students become more scientifically literate about an important global issue. In broader terms, we want to suggest ways you can use the topic of climate change to demonstrate the nature of science.

Part I, "The Science of Climate Change," includes four chapters of background information on climate and how it is changing. The chapters in Part I are suitable as reference material for busy instructors who just want "the basics" or as classroom readings for advanced high school students or nonscience majors. We have broken each chapter into short, relatively self-contained sections to make it easier to find specific topics. Chapter 1 introduces climate (including the natural greenhouse effect) and its importance to life on Earth. Chapter 2 outlines some of the methods and evidence used by scientists to detect and explain climate change. We focus on changes that have already occurred, rather than predictions of future change. Chapter 3 is an overview of the biological effects of climate change, from the responses of individual organisms to those of entire ecosystems. Chapters 2 and 3 also contain sidebars (Nature of Science boxes) about the nature of science that can help you integrate scientific processes and results in your classroom discussions about climate change. Chapter 4 is a student-friendly overview of the concepts in the first three chapters.

Part II, "Climate Change Case Studies" includes six classroom investigations for use in high school or college-level science courses. Each investigation is a case study of a well-documented biological response to climate change. Students solve real-life scientific problems using guiding questions, data tables and graphs, short reading assignments, and/or independent research. To emphasize the cooperative nature of science, all of the investigations require group work. Chapters 5 through 10 are each organized as follows:

- The introductory "At a Glance" comments give you a quick overview of the investigation.
- The teacher pages, which begin each chapter, include background information and teaching notes (i.e., materials, procedure, assessment, and extensions) specific to the investigation.
- The student pages, which follow the teacher pages, include copy-ready data sets, readings, and worksheets.

You can use as many or as few investigations as you want, and in any order. The table on page xiv can help you choose those that are most appropriate for your classroom. If you teach high school, you also can find how each activity addresses the National Science Education Standards (NRC 1996) in Connections to the Standards, page 74.

Key terms are defined in the Glossary. They appear in **boldface** at their first mention in the background text. In Chapter 4, and the student pages of Part II, especially important key terms appear in boldface a second time.

The study of how climate and its natural cycles affect biological systems is more than 150 years old. The study of how *human-dominated* climate affects biological systems is relatively new. The vast majority of scientific information on this topic has been published only since 2003 and so is not available in popular books and news articles. Most of our references, therefore, are from the primary literature (i.e., articles from scientific journals) and the Fourth Assessment Report of the International Panel on Climate Change (Parry et al. 2007; Solomon et al. 2007).

There is broad consensus among scientists on how the climate is changing and why. We regularly monitored the scientific literature and a range of climate change blogs (e.g., *http://blogs.nature.com/ climatefeedback; www.talkclimatechange.com; www.globalwarming. org*) to stay abreast of scientific developments and arguments "for" and "against" human-dominated climate change. Although a few scientists are not convinced by the data or analyses presented by their colleagues, in writing this book we have attempted to provide you and your students with the most up-to-date, mainstream science available. As we write these words, however, a scientist somewhere in the world is revising or adding to the larger body of knowledge about climate change. It is our hope that you or one of your students will contribute to that body of knowledge in the future.

REFERENCES

National Research Council (NRC). 1996. *National science education standards.* Washington, DC: National Academy Press.

Parry, M. L., O. F. Canziani, J. P. Palutikof, P. J. van der Linden, and C. E. Hanson, eds. 2007. *Climate change 2007: Impacts, adaptation and vulnerability. Contribution of Working Group II to the Fourth Assessment Report of the Intergovernmental Panel on Climate Change.* Cambridge, UK: Cambridge University Press.

Solomon, S., D. Qin, M. Manning, Z. Chen, M. Marquis, K. B. Averyt, M. Tignor, and H. L. Miller, eds. 2007. *Climate change 2007: The physical science basis. Contribution of Working Group I to the Fourth Assessment Report of the Intergovernmental Panel on Climate Change.* Cambridge, UK: Cambridge University Press.

Climate Change Case Studies (Chapters 5–10): Their Focus, Use, and Curriculum Connections

	Chapter 5 "Now You 'Sea' Ice, Now You Don't"	Chapter 6 "Population Peril"	Chapter 7 "Carrion: It's What's for Dinner"	Chapter 8 "Right Place, Wrong Time"	Chapter 9 "Ah-Choo!"	Chapter 10 "Cruel, Cruel Summer"
Focal Organism and Location	Adélie penguins, Antarctica	Polar bears, Canadian Arctic	Wolves, Wyoming, USA	Pied Flycatchers, Spain	Humans (allergies), Northern Hemisphere	Humans (heat wave mortality), Pole to Pole
Classroom Time (hours)	1–3.5	1.5–3.5	1–2.5	1.5–3	Minimum 2–3.5	Minimum 3–4
Relative Difficulty Level (1 = easiest; 3 = hardest)	1	1	2	2	3	3
Science Connections						
Scientific Process Skills	♦	♦	♦	♦	♦	♦
Ecology/ Evolution	♦	♦	♦	♦		
Environmental/ Earth Science	♦	♦	♦	♦	♦	♦
Human Health					♦	♦
Interdisciplinary Connections						
Mathematics/ Statistics			♦	♦		♦
Technology/ Computers					♦	♦
Social Studies	♦	♦	♦		♦	♦
Language/ Visual Arts		♦			♦	

About the Authors

Juanita Constible is a coastal Louisiana outreach coordinator with the National Wildlife Federation in Baton Rouge, Louisiana. After graduating with her MS in biology from the University of Victoria, Constible worked as a wildlife biologist in California, North Dakota, and Louisiana. She then spent three years as a laboratory and outreach coordinator at Miami University, in Oxford, Ohio. In that position, she coordinated professional development activities and an Antarctic outreach program for K–12 science teachers. She is the author of six research articles in scientific journals and seven articles in science education journals. This is her first book.

Luke Sandro received his MAT in biological science education from Miami University after exploring various other fields. He has been teaching biology at Springboro High School, in Ohio, for seven years. He loves teaching adolescents because it is never boring and because he gets slightly better at it every year. He has enjoyed working with Lee and Constible on research and educational outreach for several years and has learned much about science and writing from both of them. He has authored three research articles about insect cryobiology and five science education articles.

Richard E. Lee, Jr., is distinguished professor of zoology at Miami University. He received his BA from the College of Wooster (1973), and an MS (1976) and PhD (1979) from the University of Minnesota. His research focuses on physiological and ecological mechanisms of cold tolerance, dormancy, and the winter ecology of temperate and polar insects and other ectotherms. This research includes field work on Ellesmere Island in the High Arctic and five field seasons on the Antarctic Peninsula. Lee has published more than 200 articles, reviews, and book chapters and is senior editor of two books. He currently teaches courses in general entomology, winter biology, and environmental science for elementary teachers. Lee also is active in providing professional development opportunities for teachers. For the past 15 years, he has co-directed an environmental science program at a field station in Wyoming for more than 1,200 Ohio elementary teachers.

Acknowledgments

This project was supported in part by National Science Foundation grants OPP-0337656 and IBN-0416720, and grants from the Improving Teacher Quality Program administered by the Ohio Board of Regents. Any opinions, findings, and conclusions or recommendations expressed in this material are those of the authors and do not necessarily reflect those of the National Science Foundation or the Ohio Board of Regents.

Many, many thanks to the scientists and educators who critically read portions of the manuscript or piloted the classroom activities (listed alphabetically by last name):

Adriane Carlson, Talawanda High School, Ohio
Evelyn Dietz, Ursuline Academy of Cincinnati, Ohio
Amy Ferland, Greenwich High School, Connecticut
William Fraser, Polar Oceans Research Group, Montana, and
    Palmer Long-Term Ecological Research Project, Antarctica
Nate Guerin, Talawanda High School, Ohio
Emily Jones, North Central Charter Essential School, Massachusetts
Marianne Kaput, Troy Intermediate School, Ohio
Marie King, University of North Texas, Texas
Carol Landis, Ohio State University, Ohio
Marcia Lee, Miami University, Ohio
Holly LeGrand, Guntersville-Tims Ford Watershed Team,
    Alabama
Heather Marshall, Springboro High School, Ohio
Ellen Mosley-Thompson, Ohio State University, Ohio
Sheri Nuttall, Hamilton High School, Ohio
Andrew J. Petto, University of Wisconsin-Milwaukee, Wisconsin
Aaron Roberts, University of North Texas, Texas
Kathey Roberts, Lakeside High School, Arkansas
David Russell, Miami University, Ohio
Juan José Sanz, El Museo Nacional de Ciencias Naturales, Spain
Glen Schulte, Cincinnati Zoo Academy, Ohio
Doug Smith, National Park Service, Wyoming
Ian Stirling, Environment Canada, Alberta
Amanda Tolle, Miami University, Ohio
Theresa Waltner, Our Lady of the Visitation, Ohio
Chris Wilmers, University of California, Santa Cruz, California
Sarah Wise, University of Colorado at Boulder, Colorado

Photographs were donated by Lauren Danner, Michael Elnitsky, Lara Gibson, Marianne Kaput, Ashley Richmond,

Juan José Sanz, Ed Soldo, Ed Thomas, Thomas Waite, and the Yellowstone Wolf Project.

We thank Sam Bugg, Katie Meister, Celeste Nicholas, Ashley Richmond, Mandy Salminen, Amanda Tolle, and Sara Waits for obtaining research material from the library, entering data, assisting with figures, and providing other logistical support.

We greatly appreciate help from Ken Roberts, Robin Allan, Claire Reinburg, and especially Judy Cusick at NSTA Press. We also thank the review panels that considered our book proposal and completed manuscript. Their constructive suggestions made this a better book.

Juanita Constible would like to thank her husband, Kyle Haynes, for his loving encouragement and ability to alternate between serious science and kooky nonsense. She dedicates this book to the memory of Jennifer Hoffmann (1972–2007): scientist, environmentalist, and excellent friend.

Luke Sandro would like to thank his wife, Jan, for her love and patience, and his parents, Paul and Susan, for their unwavering support through difficult years.

Richard E. Lee, Jr., would like to thank his wife, Marcia, for all her love and support through the years.

Part I: The Science of Climate Change

INTRODUCTION

The average human finger has 150 ridges of raised skin that form a looped, arched, or whorled pattern. The pattern develops before birth, is unique to individuals, and is permanent throughout life. Fingerprints—impressions left by the skin ridges—can be associated with a specific person based on multiple lines of evidence. Analysts quantify individual ridge characteristics such as width, terminal branching patterns, and scarring, but ultimately use the overall order and arrangement of the characteristics to make a match (Tilstone, Savage, and Clark 2006).

Studying the biological effects of climate change is a little like fingerprint analysis. Scientists around the world have examined how organisms are affected by warming temperatures, changing precipitation regimes, and the formation of novel ecosystems. Alone, the studies are interesting but do not offer conclusive evidence that climate change is having an effect. Together, the studies strongly suggest that climate change has left a "fingerprint" on Earth's ecosystems (Parmesan and Yohe 2003).

Throughout this book we use the phrase "climate change" instead of "global warming." Although the two phrases are used interchangeably in casual conversation, they are not synonymous. **Global warming** is an increase in mean global temperatures that can contribute to changes in other climate elements. **Climate change**, on the other hand, is a significant and persistent change in *any* climate element (e.g., precipitation or clouds), at any spatial scale (e.g., regional, continental, or global) (Dow and Downing 2006).

In this section of the book, we define climate and discuss how it affects the distribution and arrangement of species and communities. We then present lines of evidence used to detect and attribute causes of climate change. Finally, we examine the overall fingerprint left by climate change on biological systems.

Topics: Global warming and global climate change
Go to: www.scilinks.org
Code: CCPPO2

REFERENCES

Dow, K., and T. Downing. 2006. *The atlas of climate change: Mapping the world's greatest challenge.* Berkeley, CA: University of California Press.

Parmesan, C., and G. Yohe. 2003. A globally coherent fingerprint of climate change impacts across natural systems. *Nature* 421(6918): 37–42.

Tilstone, W. J., K. A. Savage, and L. A. Clark. 2006. *Forensic science: An encyclopedia of history, methods, and techniques.* Santa Barbara, CA: ABC-CLIO.

Chapter 1
Climate and Life

CLIMATE BASICS

Climate is the state of the atmosphere over years or decades. Although *climate* is commonly defined as "average weather," the term encompasses more than a simple mean. It also refers to variability, seasonality, and extremes in climate elements such as temperature and precipitation (Hartmann 1994). Earth's climate is controlled by a complex, interactive system composed of land, water, snow and ice, organisms, and the atmosphere (Landsberg and Oliver 2005).

Topic: Radiation
Go to: *www.scilinks.org*
Code: CCPPO3

Earth's Energy Balance

The Sun powers Earth's climate system. Although the Sun emits **radiation** across the electromagnetic spectrum, the bulk of solar energy is visible, or **shortwave**, radiation (0.4–0.7 microns) (McArthur 2005). A fundamental property of the physical world is that the energy of a system remains constant. On an annual and global basis, solar energy entering Earth's climate system is balanced by reflection of shortwave radiation and eventual emission of **longwave**, or **far infrared, radiation** (4–100 microns) (Figure 1.1, p. 6). About 30% of incoming solar energy is reflected back to space by clouds, particles in the atmosphere, or light-colored substances on Earth's surface. The remaining 70% of incoming solar energy is absorbed by the climate system and eventually emitted back into space as longwave radiation (Landsberg and Oliver 2005).

Figure 1.1

Energy balance of Earth's climate system. The solid white arrows represent shortwave radiation emitted by the Sun. The solid gray arrows represent longwave radiation emitted by Earth's surface and atmosphere. The stippled arrow represents sensible and latent heat transfer (see "Global Circulation," p. 7). The numbers in the head of each arrow refer to the energy equivalent of a portion of solar radiation entering the atmosphere. Incoming solar radiation (100%) is balanced by the release of both shortwave (31%)

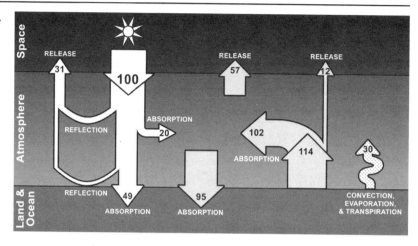

and longwave (57%+12%) radiation. However, because energy is recycled between Earth and the atmosphere (see "The Natural Greenhouse Effect," below), the land and oceans emit the equivalent of 114% of incoming solar radiation.

Source: Course S250_3, The Open University, *http://openlearn.open.ac.uk/course/view.php?id=2805.* © 2007, The Open University. Modified with permission.

Topic: Greenhouse gases
Go to: *www.scilinks.org*
Code: CCPP04

The Natural Greenhouse Effect

Solar radiation travels unimpeded through space until it encounters matter such as dust or gas. When radiation encounters matter, it may change direction without a loss of energy (scattering or reflection), pass through the matter unchanged (transmission), or be retained by the matter (absorption). The behavior of radiation depends on the physical characteristics of the matter and the wavelength of the radiation itself (Mills 2005).

Each gas in the atmosphere has a different absorption profile for short- and longwave radiation. In broad terms, transparent gases transmit radiation and opaque gases absorb it. The dominant constituents of the atmosphere—oxygen, nitrogen, and argon—are transparent to most incoming shortwave and most outgoing longwave radiation. **Greenhouse gases**, such as water vapor and carbon dioxide, are relatively transparent to shortwave radiation but opaque to longwave radiation. This selective absorption of radiation causes the natural **greenhouse effect** (Hartmann 1994).

The greenhouse effect is a process by which radiation is "recycled" between Earth and the atmosphere. Greenhouse gases in the atmosphere absorb about 20% of incoming solar radiation and more than 90% of longwave radiation emitted by the planet's surface (Figure 1.1). Much of the absorbed energy is later emitted downward, again as longwave radiation (Mills 2005). The atmosphere and the Sun actually heat Earth in roughly equal measure. Without the greenhouse effect, Earth's mean temperature would be about 33°C cooler than it is now (Pitman 2005). It is important to note that the atmosphere slows the loss of energy to space but does not *permanently* trap energy.

The term *greenhouse effect* is not completely accurate because a greenhouse doesn't heat air the same way the atmosphere heats Earth. Greenhouse gases mainly warm the atmosphere by delaying the loss of outgoing longwave radiation to space. Glass greenhouses, by contrast, mainly warm air by preventing **convection** (i.e., mixing of warm air inside the greenhouse with cold air outside the greenhouse) (Allaby 1996). As we will see in the next section, air in the atmosphere is constantly on the move.

Topic: Greenhouse effect
Go to: *www.scilinks.org*
Code: CCPPO5

Global Circulation

Although the loss and gain of radiation is balanced over the entire climate system, no one part of the planet's surface is in equilibrium at a given time (Lockwood 2005). Different areas of Earth's surface receive unequal amounts of solar radiation for a number of reasons. First, Earth is an oblate spheroid (a slightly flattened sphere). Solar radiation reaches the planet in essentially parallel lines, so it can only intersect the surface at a 90° angle near the equator. (The precise intersection varies between 23.5°N and 23.5°S over the year because of Earth's tilt and orbit.) The angle of solar rays becomes increasingly oblique at higher latitudes, reducing the intensity of radiation reaching the surface. Because rays near the poles are entering the atmosphere obliquely, they also have to travel farther through the atmosphere to get to the surface and are more likely to be absorbed by atmospheric gases (Trapasso 2005).

Second, Earth rotates on a tilted axis as it orbits the Sun. The Northern Hemisphere is tilted toward the Sun from March through September and away from the Sun the rest of the year. The hemisphere pointed toward the Sun receives more direct rays for a longer period each day (Alsop 2005).

Third, Earth's surface is composed of both liquids and solids. Water heats and cools two to three times more slowly than land because it has a high specific heat, is translucent so it can be heated more deeply, and is fluid so it can circulate heat energy (Trapasso 2005). The range of seasonal climate variability is smaller in the Southern than in the Northern Hemisphere because more of the Southern Hemisphere's surface is covered by ocean (Hartmann 1994).

Finally, some surfaces are more reflective than others. The proportion of solar energy reflected by an object, called its **albedo**, is a function of the composition, roughness, and color of that object (Figure 1.2). The albedo of soil, for example, depends on characteristics such as particle size and water content (Goward 2005). The poles, which are covered by ice and snow for much of the year, have a higher albedo than tropical regions, which are covered by dark green vegetation (Hartmann 1994).

When short- and longwave radiation gains exceed longwave losses on part of Earth's surface, the extra energy heats the air, evaporates water, or is temporarily stored by the surface (Figure 1.1). **Sensible heat transfer** occurs when energy is moved by **conduction** and convection from the surface to the atmosphere. The energy makes air molecules move faster, which we feel as an increase in temperature. **Latent heat transfer** occurs when water changes phase between a liquid, gas, or solid. When water on Earth's surface evaporates, the surface loses energy and cools—the planet is essentially "sweating." When vapor condenses in the atmosphere, the energy gained during evaporation is released and the air becomes warmer. The amount of

Topic: Albedo
Go to: *www.scilinks.org*
Code: CCPP06

Figure 1.2

Albedo of glacial ice, conifer forest, and sagebrush.

Most landscapes are a patchwork of different surfaces, but overall, the albedo of polar regions is higher than that of equatorial regions.

Source: Data compiled from Goward 2005 and Mills 2005. Photograph courtesy of Ed Soldo.

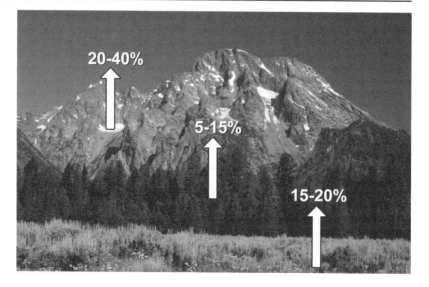

energy partitioned between heating, evaporation, and storage depends on the time of day, the availability of water, and the physical characteristics of the surface receiving the radiation (Pidwirny 2006; Ritter 2006).

Energy gradients (especially between the equator and the poles) drive the fluid motions of the atmosphere and oceans (Hartmann 1994). If our planet had a smooth, homogeneous surface and did not rotate, cells of warm air at the equator would rise, move straight north or south to the poles, and be replaced by cells of cold polar air. However, Earth's rotation causes the **Coriolis force**, which deflects wind to the right in the Northern Hemisphere and to the left in the Southern Hemisphere (Figure 1.3) (Lockwood 2005). Generally, as warm air moves poleward from the equator, it loses heat and moisture and settles near the surface at about 30°N and 30°S, the so-called **subtropical highs** (Akin 1991). On the poleward side of the subtropical highs, the **westerlies** blow air toward the poles from the southwest in the Northern Hemisphere and the northwest in the Southern Hemisphere. When the air moved by the westerlies cools and settles near the poles, relatively weak, irregular winds called the polar easterlies move air back toward the equator. On the equatorial side of the subtropical highs, the **easterlies** (also called trade winds) blow air toward the equator from the northeast in the Northern Hemisphere and the southeast in the Southern Hemisphere. The easterlies from the north and south meet in the **equatorial trough**, a belt of low pressure near the equator where winds are generally light. These global patterns of heat transport can be further distorted by the distribution of land and water and by barriers such as mountains (Akin 1991).

Figure 1.3

Simplified model of global circulation (surface winds only).

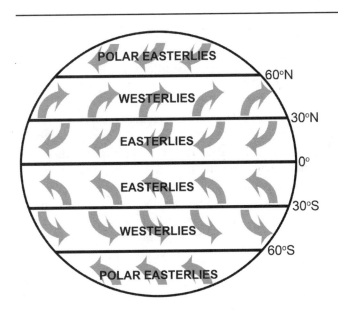

The atmosphere moves about 60% of the heat that travels from the equator to the poles. The ocean moves the remaining 40% (Hartmann 1994). Ocean currents are driven by wind and atmospheric pressure gradients at the surface and by temperature and salinity gradients below the surface. Heat also is transported vertically in the ocean by upwelling (i.e., upward flows of cold, dense water) and downwelling. Currents flowing toward the poles, like the Gulf Stream, heat the eastern coasts of continents. Currents flowing away from the poles, like the California Current, cool the western coasts of continents (Hartmann 1994).

Some regional patterns in atmospheric and oceanic circulation can impact the global climate. The most important of these patterns is the **El Niño-Southern Oscillation** (ENSO), a cycle of sea surface temperatures and atmospheric pressure that occurs in the tropical Pacific but affects temperature and precipitation worldwide. **La Niña** and **El Niño** (the cold and warm extremes of ENSO) recur every two to seven years, although El Niño events tend to be more extreme. El Niño seems to be triggered by a weakening of the easterlies and a consequent rise in sea surface temperatures in the western and central Pacific (McPhaden, Zebiak, and Glantz 2006).

Topic: La Niña/El Niño
Go to: *www.scilinks.org*
Code: CCPPO8

The Water Cycle

The three main processes that move water between Earth's surface and the atmosphere are precipitation, evaporation from the surface, and transpiration from plants. On land, evaporation and transpiration are difficult to distinguish so they are often considered together as **evapotranspiration**. Precipitation and evapotranspiration are balanced globally: All the water that enters the atmosphere eventually leaves the atmosphere. However, factors such as variations in temperature and the distribution of land alter the relative importance of these processes and, therefore, the distribution of water in time and space (Hartmann 1994).

The distribution of water is critical to the climate system in three ways. First, as mentioned above in the discussion of latent heat transfer, the energy released by phase changes of water helps drive atmospheric circulation (Hartmann 1994). Second, water vapor is the most important and abundant greenhouse gas; in clear skies, it's responsible for about 60% of the natural greenhouse effect (Trenberth et al. 2007). This is because unlike other greenhouse gases, water vapor absorbs longwave radiation across the infrared spectrum instead of in a narrow band. Finally, water changes the albedo of the planet's surface through the deposition of snow and ice, distribution of vegetation, and structure of soils (Hartmann 1994).

Climate Classification

Climate can be classified in a number of ways. The simplest methods use a single physical variable such as latitude, average temperature, or variation in the balance between precipitation and evapotranspiration. Life scientists prefer to classify climate as a combination of physical and biological elements. One common, albeit not completely standardized, scheme is **biomes** (sometimes called

Figure 1.4

(a) Alpine tundra

(b) Tropical rainforest

(a) Alpine tundra (Rendezvous Mountain, Wyoming)
(b) Tropical rainforest (Amazon Conservatory, Peru)

Photo (a) courtesy of Juanita Constible.
Photo (b) courtesy of Ashley Richmond.

zones in aquatic habitats). Biomes are defined by unique functional (i.e., morphological or physiological) groups of plants and animals adapted to a particular regional climate (Cox and Moore 2000). **Tundra**, for example, is a highly seasonal biome found around the Arctic Circle, on sub-Antarctic islands, and at high altitudes (Figure 1.4a). Tundra vegetation is typically low-lying and tolerant of extended periods of drought and cold. **Tropical rainforest**, on the other hand, is found near the equator and is characterized by hot, wet, and relatively stable conditions (Figure 1.4b). The vegetation in this biome is productive, diverse, and structurally complex (Begon, Townsend, and Harper 2006). Unless biomes are demarcated by a geographic barrier such as a mountain or a coastline, they tend to grade into one another (Cox and Moore 2000).

Topic: Biomes
Go to: www.scilinks.org
Code: CCPP09

THE BIOLOGICAL ROLE OF CLIMATE

Climate is a major force shaping the distribution, abundance, and structure of life at every level of organization, from subcellular structures to biomes.

Physiological Limits

An abiotic **limiting factor** is a physical factor that constrains an organism's life processes (Figure 1.5). Within the optimal range of a limiting factor, growth is maximized and reproduction can occur. Outside the optimal range, growth is slower and reproduction ceases. Beyond the upper and lower limits of tolerance, growth ceases and the organism may suffer irreversible damage or death (Begon, Townsend, and Harper 2006).

Figure 1.5

Generalized model of effect of limiting factors on organismal survival, growth, and reproduction.

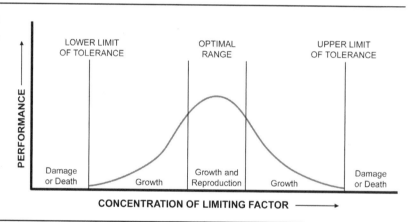

Temperature is the most important limiting factor imposed by climate because it affects nearly every aspect of an organism's physiology. At the most basic level, temperature affects how fast molecules move and, therefore, the rate of biochemical reactions (Hochachka and Somero 2002). When temperatures are too low, reactions may proceed too slowly to support life. When temperatures are too high, the pace of reactions may outstrip the availability of energy or chemical substrates (Allaby 1996).

Temperature also affects the structure and function of cellular components. Extreme heat or cold can irreversibly damage enzymes

National Science Teachers Association

and membranes and disrupt coordination among interdependent biochemical reactions (Levinton 2001). In cases of sublethal temperature stress, organisms may have to use extra energy to repair damaged cells or tissues (Hochachka and Somero 2002).

Finally, temperature can alter the water balance of an organism's cells and tissues. Water is critical to life. It is the primary constituent of organisms and acts as a chemical reagent, coolant, and transport mechanism (Akin 1991). The rate of evaporation—and therefore water loss—depends on both the temperature and relative humidity of the surrounding air. Organisms lose more water in extreme heat. They also lose more water in extreme cold, because the water content of the air is lower than that of their bodies (Begon, Townsend, and Harper 2006).

Precipitation is another critical limiting factor controlled by climate. Precipitation both directly affects the availability of water and indirectly affects the intensity of other physical limiting factors. For example, precipitation can change the salinity, sediment load, and nutrient levels of coastal systems by increasing discharge from rivers (Kaiser et al. 2005). Furthermore, precipitation can change the pH of streams and lakes via acid rain (Begon, Townsend, and Harper 2006).

The distribution and abundance of a species depends in part on the physiological limits of its individuals (Spicer and Gaston 1999). Consider, for example, the mosquito *Anopheles gambiae sensu stricto*, a major vector of malaria (Figure 1.6). The potential geographic range of this species is defined by both temperature (5–42°C) and precipitation (33–332 cm per year). The abundance of the species, however, varies across its range depending on how favorable the conditions are. Although the mosquito *can* develop outside its optimal range, it is most abundant where conditions are warm and wet (Figure 1.6) (Lindsay et al. 1998).

Figure 1.6

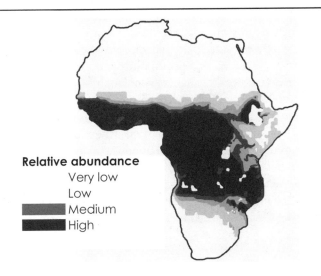

Relative abundance
Very low
Low
Medium
High

Approximate distribution and relative abundance of *Anopheles gambiae sensu stricto* in Africa.

Source: Lindsay, S. W., L. Parson, and C. J. Thomas. 1998. Modified with permission.

Evolution

No one species can withstand the full range of physical conditions found on Earth (Spicer and Gaston 1999). In part, this is because adaptive evolution can occur only in response to local conditions (Begon, Townsend, and Harper 2006). Most organisms live within **microclimates**, small areas that experience significantly different physical conditions than those of the regional climate or **macroclimate** (Figure 1.7). Microclimates vary in size, from a tiny bubble of air to an entire hillside or valley, and are formed by variations in geometric position, shading, and other physical and biological factors (Bailey 2005).

Figure 1.7

Microclimates in a valley on Santa Catalina Island, California.
The relatively cool and wet north-facing slope of the valley (left side of the picture) is dominated by oak trees and a variety of herbs. The relatively warm and dry south-facing slope is dominated by grasses and drought-resistant plants such as sagebrush and cactus.

Photograph courtesy of Lauren Danner.

Earth's climate is always changing because of astronomical, geological, and biological processes. If a particular change is correlated with variation in reproductive success, a **population** (group of individuals of one species) may adapt to that change. Adaptation to a new climate regime can occur relatively rapidly (i.e., a few generations to a few hundred years) in large, genetically variable populations (Stearns and Hoekstra 2005). Climatic change also can trigger the evolution of new species. Both extreme events and gradual trends over time can increase the chance of speciation by isolating populations, forcing populations to relocate, or presenting new selection pressures. In mammals, at least, it seems climatic changes must be unusual over a period of hundreds of thousands of years to cause the evolution of new species (Barnosky and Kraatz 2007).

Structure and Function of Ecosystems

An individual organism—no matter how solitary its lifestyle—constantly interacts with other organisms. Within populations, individuals may compete for resources, prey on one another, or cooperate to raise young. The strength of interactions within and between populations helps shape the structure and function of **communities** (groups of populations of different species) (Begon, Townsend, and Harper 2006).

The relationship between climate and the structure and function of communities is complicated. On one hand, the internal dynamics of a community may override the climatic suitability of a habitat for a given species. In New England, for instance, the northern limit of the little gray barnacle (*Chthamalus fragilis*) is set by competition, not climate. In the absence of competition from northern acorn barnacles (*Semibalanus balanoides*), little gray barnacles could live 80 km farther north than they currently do (Wethey 2002). On the other hand, climate mediates species interactions—for example, by controlling the rate at which energy flows through a food web. Seasonal and annual variations in temperature, precipitation, and the number of cloudless days regulate the amount of new tissue produced by plants (Begon, Townsend, and Harper 2006). Climate also affects interactions at the top of the food chain. On the western coast of North America, the ochre sea star (*Pisaster ochraceus*) consumes more of its primary prey (i.e., mussels) when temperatures are high. Under consistently high temperatures, sea stars can extensively damage mussel beds, eliminating the habitat of hundreds of intertidal species (Harley et al. 2006).

Another layer of complexity is added when we consider the relationship between **ecosystems** (communities and their associated physical environments) and climate. Climate influences local soil types, decomposition rates, and the turnover of nutrients between soils and living organisms. Ecosystems also affect local climate, however, through their surface properties (e.g., albedo) and exchanges of energy, water, and gases. For example, crops and grasses can enhance the effects of wetter-than-average summers. In the Mississippi River Basin of North America, increased soil moisture leads to increased plant growth. More plants mean more leaf surface area and, in turn, higher evapotranspiration rates. The rapid cycling of water from the soil to the atmosphere increases water vapor and, therefore, precipitation (Kim and Wang 2007). The cyclic magnification of a regional climate trend is called a **feedback**. We will discuss feedbacks further in Chapter 2.

THE ENHANCED GREENHOUSE EFFECT

Humans have had a significant impact on Earth's climate system despite its size and complexity. The natural greenhouse effect, which keeps the planet livable, has been enhanced globally by fossil fuel emissions and regionally by feedbacks. In the next chapter, we will discuss how scientists study climate change. We also will examine the types and causes of changes that have occured in the last 100 years.

REFERENCES

Akin, W. E. 1991. *Global patterns: Climate, vegetation, and soils.* Norman, OK: University of Oklahoma Press.

Allaby, M. 1996. *Basics of environmental science.* London: Routledge.

Alsop, T. 2005. Seasons. In *Encyclopedia of world climatology*, ed. J. E. Oliver, 651–655. Dordrecht, The Netherlands: Springer.

Bailey, W. G. 2005. Microclimatology. In *Encyclopedia of world climatology*, ed. J. E. Oliver, 486–500. Dordrecht, The Netherlands: Springer.

Barnosky, A. D., and B. P. Kraatz. 2007. The role of climatic change in the evolution of mammals. *BioScience* 57(6): 523–532.

Begon, M., C. R. Townsend, and J. L. Harper. 2006. *Ecology: From individuals to ecosystems.* 4th ed. Malden, MA: Blackwell Publishing.

Cox, C. B., and P. D. Moore. 2000. *Biogeography: An ecological and evolutionary approach.* 6th ed. Oxford: Blackwell Science.

Goward, S. N. 2005. Albedo and reflectivity. In *Encyclopedia of world climatology*, ed. J. E. Oliver, 32–35. Dordrecht, The Netherlands: Springer.

Harley, C. D. G., A. R. Hughes, K. M. Hultgren, B. G. Miner, C. J. B. Sorte, C. S. Thornber, L. F. Rodriguez, L. Tomanek, and S. L. Williams. 2006. The impacts of climate change in coastal marine systems. *Ecology Letters* 9(2): 228–241.

Hartmann, D. L. 1994. *Global physical climatology.* San Diego, CA: Academic Press.

Hochachka, P. W., and G. N. Somero. 2002. *Biochemical adaptation: Mechanism and process in physiological evolution.* New York: Oxford University Press.

Kaiser, M. J., M. J. Attrill, S. Jennings, D. N. Thomas, D. K. A. Barnes, A. S. Brierley, N. V. C. Polunin, D. G. Raffaelli, and P. J. Le B. Williams. 2005. *Marine ecology: Processes, systems, and impacts.* Oxford: Oxford University Press.

Kim, Y., and G. Wang. 2007. Impact of vegetation feedback on the response of precipitation to antecedent soil moisture anomalies over North America. *Journal of Hydrometeorology* 8(3): 534–550.

Landsberg, H. E., and J. E. Oliver. 2005. Climatology. In *Encyclopedia of world climatology*, ed. J. E. Oliver, 272–283. Dordrecht, The Netherlands: Springer.

Levinton, J. S. 2001. *Marine biology: Function, biodiversity, ecology.* 2nd ed. New York: Oxford University Press.

Lindsay, S. W., L. Parson, and C. J. Thomas. 1998. Mapping the ranges and relative abundance of the two principal African malaria vectors, *Anopheles gambiae sensu stricto* and *An. arabiensis,* using climate data. *Proceedings of the Royal Society of London* B 265(1399): 847–854.

Lockwood, J. G. 2005. Atmospheric circulation, global. In *Encyclopedia of world climatology*, ed. J. E. Oliver, 126–134. Dordrecht, The Netherlands: Springer.

McArthur, L. J. B. 2005. Solar radiation. In *Encyclopedia of world climatology*, ed. J. E. Oliver, 667–673. Dordrecht, The Netherlands: Springer.

McPhaden, M. J., S. E. Zebiak, and M. H. Glantz. 2006. ENSO as an integrating concept in earth science. *Science* 314(5806): 1740–1745.

Mills, G. 2005. Radiation climatology. In *Encyclopedia of world climatology*, ed. J. E. Oliver, 603–611. Dordrecht, The Netherlands: Springer.

Pidwirny, M. 2006. *Fundamentals of physical geography.* 2nd ed. Retrieved October 4, 2007, from *www.physicalgeography.net/fundamentals/contents.html*

Pitman, A. J. 2005. Greenhouse effect and greenhouse gases. In *Encyclopedia of world climatology*, ed. J. E. Oliver, 391–397. Dordrecht, The Netherlands: Springer.

Ritter, M. E. 2006. *The physical environment: An introduction to physical geography.* Retrieved October 4, 2007, from *www.uwsp.edu/geo/faculty/ritter/geog101/textbook/title_page.html*

Spicer, J. I., and K. J. Gaston. 1999. *Physiological diversity and its ecological implications.* Oxford: Blackwell Science.

Stearns, S. C., and R. F. Hoekstra. 2005. *Evolution: An introduction.* 2nd ed. New York: Oxford University Press.

Trapasso, L. M. 2005. Temperature distribution. In *Encyclopedia of world climatology*, ed. J. E. Oliver, 711–716. Dordrecht, The Netherlands: Springer.

Trenberth, K. E., P. D. Jones, P. Ambenje, R. Bojariu, D. Easterling, A. Klein Tank, D. Parker, F. Rahimzadeh, J. A. Renwick, M. Rusticucci, B. Soden, and P. Zhai. 2007. Observations: Surface and atmospheric climate change. In *Climate change 2007: The physical science basis. Contribution of Working Group I to the Fourth Assessment Report of the Intergovernmental Panel on Climate Change,* eds. S. Solomon, D. Qin, M. Manning, Z. Chen, M. Marquis, K. B. Averyt, M. Tignor, and H. L. Miller, 235–336. Cambridge: Cambridge University Press.

Wethey, D. S. 2002. Biogeography, competition, and microclimate: The barnacle *Chthamalus fragilis* in New England. *Integrative and Comparative Biology* 42(4): 872–880.

Chapter 2
Earth's Changing Climate

HOW IS CLIMATE CHANGE DETECTED?

In 1896, Svante Arrhenius published the first **model** of the effects of industrial carbon dioxide (CO_2) on Earth's climate. The chemist's model included tens of thousands of pencil-and-paper calculations but was actually fairly simple. Arrhenius accounted for Earth's energy balance (see Chapter 1) and crude water vapor feedbacks, but ignored winds, oceans, and clouds (Weart 2003). He included some experimental data but lacked global observations of the climate system (Bowen 2005).

Since the days of Arrhenius, scientists have moved from pencils to supercomputers. Calculations take hours or days instead of months. Models include a dazzling array of variables, including evapotranspiration and oceanic heat transport. And perhaps most remarkable, scientists have data from multiple millennia and nearly every corner of the planet (NOS Box 2.1).

Scientists interested in climate change can perform small-scale experiments on, for example, the atmospheric lifetime of greenhouse gases (Dalmasso et al. 2006). It is neither ethical nor possible, however, to perform controlled experiments on Earth's entire climate system. Instead, scientists investigate how and why the climate changes by reconstructing the past, observing the present, and modeling the future.

Nature of Science (NOS) Box 2.1: Scientific knowledge is both tentative and durable.

Explanations for natural phenomena can be modified by new observations and theories but are built on previous knowledge. Modifying a scientific explanation is similar to repairing a car. If you had trouble with your windshield wipers, you wouldn't replace your engine block! Likewise, scientists usually make small "repairs" to an explanation based on observations or experimental data; rarely is it necessary to replace the theories at the explanation's core.

Today's climate models are complex and high-tech but still share basic principles of physics, chemistry, and mathematics with earlier models. Because different generations of models share the same scientific core, they also may have similar results. For example, one of Arrhenius's most important calculations was climate sensitivity, or how much warmer Earth will be on average if the pre-industrial concentration of CO_2 is doubled. Arrhenius estimated a sensitivity of 5–6°C (Bowen 2005), just 2–3°C higher than the current best estimate (IPCC 2007a). Note that despite the fact that results of sensitivity modeling have been remarkably similar, the estimates are still quite uncertain. Scientists will continue to adjust sensitivity models as new research methods and tools are developed.

The Past

An understanding of past climates helps us interpret the present and forecast the future. The **instrumental record** (discussed in the next section) describes only the last 150 years of Earth's climate. **Paleoclimatologists**, on the other hand, can use **historical sources** and **proxies** to peek into millions of years of Earth's past (Jansen et al. 2007).

Historical sources are human-made records such as cave paintings, crop harvest diaries, and ship logs. These sorts of records are limited in their geographic and temporal scope. They also can suffer inaccuracies because of issues such as changes in language. For example, before about 1750, English mariners recorded nearly every wind above 21 km/hr as a "gale" (Jones and Mann 2004). Today, *gale* only refers to winds between 51 and 101 km/hr (AMS 2000).

Proxies are physical, biological, or chemical features that indirectly reflect changes in climate elements such as temperature (Table 2.1). For example, paleoclimatologists can infer past conditions by comparing the geographic distribution and abundance of pollen preserved in lake sediments with their modern counterparts (Jansen et al. 2007). Proxies vary in their spatial coverage, their length of record (i.e., how much time in Earth's history they describe), and their temporal resolution (Table 2.1). Terrestrial boreholes, for instance, span up to 1,000 years and can describe only multidecadal to centennial variations in climate. The tree ring record, on the other hand, spans up to 10,000 years and can describe decadal or even annual variations in climate (Jones and Mann 2004). Proxies also vary in their reliability. Generally, the further back in time a record extends, the more complicated it can be to interpret. Paleoclimatologists can minimize the shortcomings of individual methods by using multiple proxies (Jansen et al. 2007).

Table 2.1

Examples of proxies commonly used by paleoclimatologists to reconstruct past climates.

Proxy	Climate information	Spatial coverage	Length of record (years ago)
Fossil corals	Temperature, salinity	Tropical and subtropical oceans	100–50,000
Terrestrial boreholes	Temperature	Global	300–1,000
Fossil pollen	Temperature, precipitation	Global	500–100,000
Tree rings	Temperature, precipitation, solar radiation	Subpolar and mid-latitude terrestrial areas	10,000
Lake sediments	Temperature, precipitation	Global	10,000–100,000
Ice cores	Accumulation of precipitation, volcanic eruptions, temperature, solar radiation, drought, atmospheric chemistry and transport	Polar and alpine regions	40,000–650,000
Glacial geologic features	Former extent and timing of ice sheets	Poles to midlatitudes; high altitude alpine areas in tropics	1 million
Fossil foraminifera	Temperature, atmospheric and oceanic chemistry	Southern Ocean, South Atlantic, and Pacific	>50 million

Source: Compiled from Jansen et al. 2007; Jones and Mann 2004; Saltzman 2002.

The Present

The instrumental record consists of direct measurements of climate elements such as temperature, precipitation, humidity, wind speed and direction, and atmospheric chemistry. The record also includes indirect indicators of climate change, such as sea level (Trenberth et al. 2007).

At Earth's surface, temperature is recorded by more than 4,000 thermometers at terrestrial stations and thousands more on seafaring ships. Above the surface, measurements are taken by weather balloons and orbiting satellites. Precipitation is recorded directly by gauges or snow stakes and remotely by radar, microwave, or infrared sensors (Trenberth et al. 2007).

Scientists use infrared gas analyzers, which identify gas molecules by how they absorb longwave radiation (Palassis 1999), to measure atmospheric carbon dioxide (CO_2). Because the analyzers are expensive and difficult to operate, there are relatively few baseline stations with continuous onsite CO_2 records. The baseline stations are supplemented by an extensive network of marine and terrestrial sites; technicians collect weekly samples in flasks and send them to central laboratories for analysis (Forster et al. 2007). Note, however, that CO_2 emissions are mixed rapidly in the atmosphere, so annual increases of CO_2 differ little between the station at Barrow, Alaska, and the one at Mauna Loa, Hawaii (Thoning, Kitzis, and Crotwell 2007). You can explore CO_2 trends (changes in CO_2 over time) at the National Oceanic and Atmospheric Administration's Interactive Atmospheric Data Visualization website (*www.esrl.noaa.gov/gmd/ccgg/iadv*).

It is a massive task to maintain a real-time, global climate data set. The data must be continuously updated, checked for errors, standardized, and compiled in central repositories. This global cooperation allows scientists to detect climate change and to determine its causes (NOS Box 2.2).

Nature of Science (NOS) Box 2.2: Science is collaborative, international, and multidisciplinary.

Early scientists rarely specialized in just one discipline. Galileo, for example, was a physicist, astronomer, and mathematician. Marie Curie was a physicist and chemist. Today, however, each discipline increasingly requires its own prior knowledge, skill set, technology, and statistical techniques. Productive scientists tend to spend most of their careers on topics *within* disciplines (e.g., string theory or breast cancer).

The study of climate change includes physics, meteorology, oceanography, glaciology, ecology, biogeography, paleontology, and more. No one researcher can be an expert in all these fields, and no one country has sufficient resources to collect worldwide data sets. Few scientific enterprises involve the same degree of collaboration as the study of global climate change.

The Future

A mathematical model is a numerical representation of a complex system. Natural and social scientists use models to define important relationships in a system and to make realistic projections about the future. Models are not games with predetermined outcomes—they follow the same guidelines as other forms of scientific inquiry.

First, models are based on prior scientific knowledge (NOS Box 2.1). Climate models must follow and accurately quantify physical laws such as conservation of mass. They also increasingly incorporate information on complex climate features such as feedback loops (see "Feedbacks," pp. 39–40). Second, models must be able to simulate real-world observations. Scientists routinely test the performance of their models by comparing simulations of the climate with the real thing (Figure 2.9, p. 38). In the past decade, models have shown increasing accuracy in representing patterns of climate variability such as seasonal changes in sea ice. Finally, models must be evaluated by many scientists (NOS Box 2.3). For instance, different research groups might methodically compare their "competing" models to identify biases, missing climate processes, or mechanistic flaws (Randall et al. 2007).

Nature of Science (NOS) Box 2.3: Science is a public enterprise.

Scientists are human. They can be illogical, they make mistakes, and they sometimes have ethical lapses. The public nature of science helps protect the quality and integrity of scientific inquiry. If a result rests on faulty reasoning or sloppy technique, it is openly criticized or even rejected.

Scientific results and explanations are reviewed at multiple levels. When investigators apply for research funding, they submit a study plan to a panel of experts for approval. As the study progresses, investigators often seek advice from colleagues. The results of the study might be presented at an international conference, where scientists from other institutions or disciplines can publicly comment on the work. If yet *another* panel of experts deems the completed study interesting and technically sound, it is published in a scientific journal. Finally, independent teams of scientists may, at any time, attempt to replicate the published results.

Climate models are not without uncertainties (see "Uncertainty and Consensus," p. 24). Regional processes, like changes in cloud cover, play a critical role in climate change but are difficult to represent in global models. In fact, clouds are one of the biggest reasons climate models disagree on how much the climate will warm in the future. Increased computing power and observational data will help resolve these disagreements (Le Treut et al. 2007). In the meantime, models unanimously project that increasing greenhouse gas concentrations will substantially warm the climate (Randall et al. 2007).

Uncertainty and Consensus

"Of course scientists disagree—if all scientific questions were resolved, there would be no need for scientific investigation or for scientists, other than those who write science textbooks."

—William Ascher, *"Scientific Information and Uncertainty: Challenges for the Use of Science in Policymaking,"* 2004

Uncertainty is an important—and inescapable—part of science. It identifies gaps in our knowledge and stimulates new research. In the study of climate change, there are two main types of uncertainty. **Value uncertainties** occur when data are inaccurate or incomplete (Le Treut et al. 2007). The precipitation record, for example, is both inaccurate and incomplete. Rain gauges can underestimate precipitation in windy environments, and large geographical areas (e.g., most of Africa) have been insufficiently sampled. Scientists can partially compensate for inaccurate or incomplete data with statistical techniques such as interpolation (i.e., estimation of a missing value between two known values) (Trenberth et al. 2007). **Structural uncertainties** occur when explanations or models are missing relevant processes or relationships. As discussed in the previous section, scientists have difficulty modeling the contribution of clouds to climate change because they don't fully understand how aerosols alter cloud albedo (Le Treut et al. 2007).

Despite the uncertainties in climate research, there is broad **scientific consensus** that humans are significantly altering Earth's climate. The most detailed statement of consensus has been released by the Intergovernmental Panel on Climate Change (IPCC). IPCC reports are written by teams of scientists who critically review thousands of peer-reviewed scientific studies, industry reports, and accounts of traditional knowledge. The 2007 report was compiled by more than 1,250 authors and 2,500 reviewers in more than 130 countries (IPCC 2007b). Several other scientific organizations also have issued consensus statements on human-dominated climate change (e.g., AGU 2008; Joint Science Academies 2005; AAAS 2007).

A scientific consensus is not a statement of absolute truth. It doesn't even imply that all scientists agree with all details of an explanation (Pielke 2005). Rather, a consensus recognizes that an explanation is logical, consistent with previous knowledge, and supported by multiple lines of evidence (Gauch 2003). Scientific consensus makes it possible to have things like vaccination programs, safety standards for automobiles, and physics or biology textbooks. In short, consensus creates a "base of operations" from which to plan and carry out further research.

Unlike the current consensus on climate change, there was no broad agreement on predictions of "global cooling" in the early 1970s. Based on a few lines of evidence, a small group of scientists reasoned that Earth would plunge into an ice age as pollution eventually blocked energy from the sun. The group was so concerned that they sent a letter to President Nixon and spoke publicly around the United States. Although the broader scientific community recognized that Earth's climate might be changing, it was split over whether the world would cool in the future due to particulate pollution, or warm due to CO_2 emissions (Weart 2003).

WHAT IS THE EVIDENCE FOR CLIMATE CHANGE?

"Scientists rarely accept a result until it is confirmed by wholly different means."
—Spencer Weart, *The Discovery of Global Warming*, 2003

The natural variability of Earth's climate makes it difficult to detect out-of-the-ordinary change. In the past few decades, however, scientists have collected multiple lines of evidence that consistently indicate the world is warming. Here, we present some of the evidence most relevant to biological systems.

Temperature

Since 1906, the average global surface temperature increased by about 0.7°C (Figure 2.1, p. 26), and from 1995 to 2007, Earth experienced 12 of the 13 warmest years on record since 1850 (McCarthy 2008; Trenberth et al. 2007). These statistics are *global averages*, meaning not all regions of Earth have experienced the same degree of warming. Some regions of Earth have warmed less than average—or even cooled, slightly—and some regions have warmed more than average (3–4°C in the last century).

Regional patterns in warming are related to latitude, patterns of atmospheric circulation, and many other factors. For instance, air over the land is warming faster than air over the oceans (Figure 2.1) because land can hold less heat than water (Trenberth et al. 2007). The poles tend to warm faster than the tropics, because cold, dry air is more sensitive to the addition of CO_2 and water vapor than hot, humid air (Le Treut et al. 2007). Some parts of Antarctica appear to be cooling, likely because of a large, regional "hole" in the ozone layer (ozone is a greenhouse gas) and because storms that typically occur at lower latitudes are shifting, bringing cold air farther south than usual (Trenberth et al. 2007).

Figure 2.1

Annual deviation of global air and sea surface temperatures from the long-term mean (average of all years from 1961 to 1990).

Temperatures at the dotted line are equal to the long-term mean. Warmer-than-average years appear above the line, whereas cooler-than-average years appear below the line. See NOS Box 2.4 for an explanation of how (and why) deviations are calculated.

Source: Trenberth et al. 2007. Modified with permission.

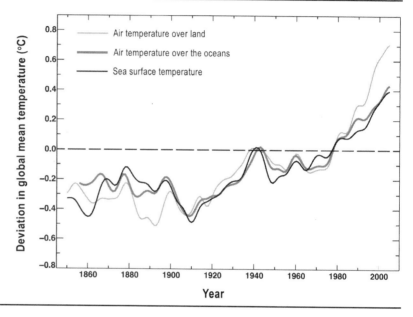

Urbanization also influences regional climate. Cities tend to experience higher temperatures than surrounding rural areas, because of changes to the land surface, air pollution, and other factors (see Chapter 10 for more details). Despite the number of cities worldwide, urbanization probably doesn't have a significant effect on global temperature. Some studies suggest that global trends are about 10 times larger than urban trends, and in the United States and China, rural sites showed warming trends that were nearly identical to trends from rural *and* urban sites. To be on the safe side, however, scientists exclude as many urban sites as possible from global temperature averages (Trenberth et al. 2007) (NOS Box 2.4).

Earth has experienced other warm periods in the past (Figure 2.2, p. 28). These warm periods, however, have been relatively short, making up just 20% of each glacial-interglacial cycle on average. Earth also has experienced other rapid global-scale changes in climate.

Between the last ice age and the current interglacial period, for example, the planet warmed by 4–7°C. However, this process took about 5,000 years (i.e., a rate 5–9 times slower than the modern warming trend) (Jansen et al. 2007).

Nature of Science (NOS) Box 2.4: Scientists work to avoid bias.

Although scientists try to be objective, anything from their personal opinions to the tools they use can introduce bias to a study. Scientists need to be mindful of potential sources of bias and how those biases may affect their work. Some forms of bias can be minimized with proactive measures such as training and careful study design. Other forms are undetectable until a study is complete, or are unavoidable.

Bias sometimes can be dealt with mathematically. To accurately calculate the absolute global temperature, for example, scientists would need to sample equally at all elevations, at all latitudes, and in all habitat types. In the real world, the number and location of weather stations is always changing because of financial considerations and the ability to gain access to remote locations. How can scientists be sure that increasing global temperatures are related to a real change in temperature instead of an increase in sampling effort in warm areas of the globe (e.g., cities)? The solution is to calculate the difference (called **climate deviation** or anomaly) between the temperature at a given time (e.g., June 1987) and weather station and the long-term average for the same station (see example below and Figure 2.1) (Jones et al. 1999). Scientists can then calculate mean global deviations for a given time by averaging the deviations from all available weather stations.

Station	Station Temperature at Time X (°C)	Mean Station Temperature Over 30 Years (°C)	Deviation (°C)
A	7	8	-1
B	16	13	+3
C	14	14	0
		Mean Global Deviation for Time X	+0.7

You may notice from the figures in this chapter that global deviations in temperature, drought, and sea level do not all use the same reference period. By international agreement, climatologists calculate long-term climate means (also called **climate normals**) over 30-year periods. When scientists were calculating the deviations shown here, they used the 30-year period that had the best available level of sampling effort.

Precipitation

Precipitation is a difficult phenomenon to summarize globally because it occurs intermittently. It also has a complex relationship with temperature, atmospheric circulation, soil moisture, and other factors. This complexity is reflected in the magnitude and direction of recent regional changes in precipitation. Some areas have become wetter, and others have become drier (Table 2.2, p. 29) (Trenberth et al. 2007).

High temperatures promote rapid evapotranspiration, moving water from Earth's surface to the atmosphere (see "The Water Cycle," Chapter 1, p. 10). As global temperatures rise, there's an increased risk of both heavy rain (due to an increase in atmospheric water vapor) and long-term drought (due to drying out of soils). Some regions such as central Europe have suffered catastrophic floods one year and record-breaking drought the next (Trenberth et al. 2007). Whereas there currently is no evidence that floods are increasing, limited data suggest the frequency and severity of droughts has increased worldwide (Figure 2.3) (Rosenzweig et al. 2007).

Figure 2.2

Variations in atmospheric CO_2, atmospheric CH_4, and deuterium (δD), as derived from air bubbles in Antarctic ice cores.

There is a linear relationship between δD, which is an isotope of hydrogen, and local temperature. The gray vertical bars indicate interglacial warm periods of the last 450,000 years.

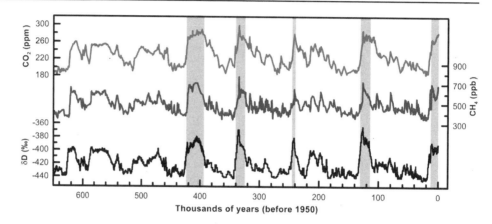

Source: Jansen et al. 2007. Modified with permission.

Figure 2.3

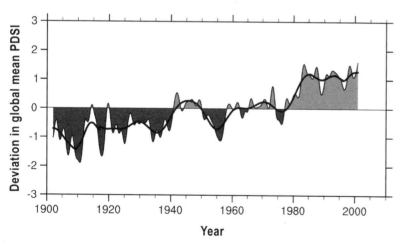

Annual deviation of the global Palmer Drought Severity Index (PDSI) from the long-term mean (average of all years from 1950–1979).

The PDSI is an index of soil moisture based on local weather patterns. Light gray areas are drier-than-average periods and dark gray areas are wetter-than-average periods. The black line is a smoothed trend in the PDSI. See NOS Box 2.4 for an explanation of how (and why) deviations are calculated.

Source: Trenberth et al. 2007. Modified with permission.

Table 2.2

Significantly more precipitation (+5 to +60% per century) • Western Canada • Southeastern Canada and northeastern United States • Eastern South America • Northern Europe • Northern and central Asia • Western Australia *Significantly less precipitation (-20 to -80% per century)* • Southwestern South America • Mediterranean • Northwestern and central Africa • Southern Asia	**Long-term trends in precipitation from 1900 to 2005.** Source: Data compiled from Trenberth et al. 2007.

Hurricanes

Hurricanes are heat engines—systems that convert heat from the surface waters of the ocean into mechanical work in the form of wind and waves (Emanuel 1987). As a hurricane develops, water is evaporated from the ocean's surface and forced upward through the storm. The vapor cools as it rises, condenses, and releases latent heat that fuels the storm's winds (Hobgood 2005).

Some studies indicate that since 1970, the number of severe hurricanes (i.e., category 4 or 5) has increased globally by 75% (Trenberth et al. 2007). However, the question of how climate change affects the number and intensity of hurricanes is still contentious. Tropical storms (i.e., systems with maximum sustained winds of at least 17 m/s) can develop into hurricanes (i.e., winds of at least 39 m/s) when sea surface temperatures are above 26°C, there is ample moisture in the atmosphere, and the difference in wind speed between the surface and the lower layer of the atmosphere is relatively low (Hobgood 2005). It would seem that increasing sea surface temperatures would automatically increase the likelihood of frequent, strong hurricanes. But a number of other factors—including the average speed of regional surface winds—may be more important. Furthermore, some leading experts caution that the quantity and quality of observational data hampers our ability to assess trends in hurricanes (Emanuel, Sundararajan, and Williams 2008).

Variation in the El Niño-Southern Oscillation (ENSO; see "Global Circulation," Chapter 1, p. 7) also adds uncertainty to the scientific understanding of hurricanes. ENSO directly influences the location and activity of tropical storms. In El Niño years, for example, hurricanes tend to be stronger and more numerous in the Pacific but weaker in the Atlantic (McPhaden, Zebiak, and Glantz 2006). The

ENSO cycle itself appears to be changing—El Niño events have become stronger and more prolonged in the last 30 years—but there's no definitive evidence that this trend is due to climate change (Trenberth et al. 2007).

The Cryosphere

The **cryosphere**, or frozen portion of Earth's surface, is a sensitive indicator of our warming climate. Across the globe, snow cover is declining, sea ice is melting, frozen ground is thawing, and glaciers are shrinking (Lemke et al. 2007). As we discuss later, loss of the cryosphere amplifies both regional and global warming (see "Feedbacks," p. 39).

When temperatures are high, more precipitation falls as rain than as snow and snowmelt is accelerated. In the Northern Hemisphere, where snow typically covers more than 33% of the land surface from November to April, spring snow cover has declined by about 2% per decade since 1966 (Lemke et al. 2007). The trend in snow cover in the Southern Hemisphere is less clear. Because the Southern Hemisphere is primarily covered by oceans, which moderate its climate, little land area outside Antarctica is covered with snow. Furthermore, few long-term records of snow cover exist for the Southern Hemisphere. The limited data available, however, suggest that some mountainous regions have experienced decreases in snow cover due to springtime warming (Lemke et al. 2007).

At both poles, climate proxies and the instrumental record indicate that sea ice is declining. In the Arctic, the extent of sea ice ranges from 7×10^6 km^2 in the summer to 16×10^6 km^2 in the winter (Serreze, Holland, and Stroeve 2007). Since 1979, the summer ice extent in the Arctic has decreased by about 10% per decade (Figure 2.4), or 72,000 km^2 per year (NSIDC 2007). In the Antarctic, the extent of sea ice ranges from 3×10^6 km^2 in the austral summer to over 19×10^6 km^2 in winter. Thus far, likely due to a lack of long-term data, scientists have not identified significant trends for the entire Antarctic continent (Lemke et al. 2007). There is, however, evidence that sea ice has declined west of the Antarctic Peninsula (see Chapter 5 for more details).

Permafrost is a layer of subsurface soil that remains frozen year-round. It covers approximately 24% of the land area in the Northern Hemisphere and plays a vital role in surface structure and associated drainage systems. There are currently no good global data on permafrost trends, but regional reports suggest significant thawing. For example, permafrost seems to be thawing up to 0.04 m/yr in Alaska and 0.02 m/yr on the Tibetan Plateau

Figure 2.4

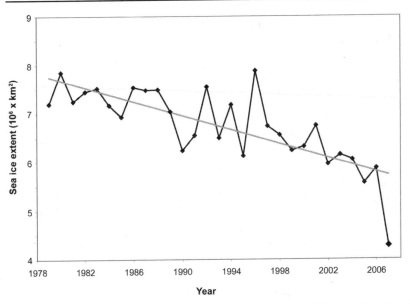

Trend in summer sea ice extent in the Arctic, 1979–2007.

Source: Figure courtesy of the National Snow and Ice Data Center.

(Lemke et al. 2007). In northern Alaska and the Northwest Territories, loss of permafrost has caused buildings, roads, and pipelines to slump (Hinzman et al. 2005).

Changes in glaciers and ice caps tend to lag behind atmospheric change by years to centuries. In spite of the lag, Earth's warming climate has caused most mountain glaciers and ice caps to shrink (Lemke et al. 2007). A glacier's annual gain and loss of snow and ice is controlled by temperature, precipitation, and how fast the glacier is flowing into the ocean (glaciers that end below the sea surface move into the water more rapidly when they start to thin) (Meier et al. 2007). Scientists estimate that from 1960 to 2004, glaciers and ice caps (excluding large ice sheets in Greenland and Antarctica) lost about 155 **gigatons** (Gt) of ice per year. (One Gt is equal to 1 trillion kilograms, or about 200 million full-grown Asian elephants.) From 1996 to 2005, the rate of ice loss was about 420 Gt/year (Meier et al. 2007). Proxies such as ice cores and fossilized wood above the modern tree line suggest that the recent rate of glacial retreat is unprecedented in at least the last 12,000 years (Jansen et al. 2007).

The Oceans

From 1955 to 1998, the salinity of the ocean decreased at high latitudes in both hemispheres, primarily because of an increase in precipitation. Other factors such as glacial melt water and increased runoff also may have contributed to this trend. Conversely, salinity increased in the shallower parts of tropical and subtropical oceans because of an increase in evaporation due to high temperatures (Bindoff et al. 2007).

The concentration of carbon has increased in the ocean, both because of increasing atmospheric CO_2 (see "Forcings," p. 33) and changes to the natural carbon cycle. When CO_2 is dissolved in water it forms a weak acid, so as the amount of carbon in the ocean increases, the pH decreases. The average pH of surface waters in the open ocean (which is typically 7.9–8.3) has decreased by 0.1 units since 1750 (Bindoff et al. 2007). Ice core data suggest that oceanic pH is at its lowest point in at least 420,000 years (Hoegh-Guldberg et al. 2007).

The average global sea level has risen by nearly 2 mm/yr since the late 1800s (Figure 2.5) (Bindoff et al. 2007). This rise in sea level is due primarily to the melting of land ice (about 60% from glaciers and ice caps alone) and secondarily to heat-induced expansion of the upper layers of the ocean (Meier et al. 2007). Although global sea level has changed dramatically before in Earth's past (e.g., a 120 m change after the last ice age), current conditions are unusual compared with the last two to three thousand years (Bindoff et al. 2007).

WHY IS CLIMATE CHANGE OCCURRING?

Changes in Earth's climate aren't, in themselves, novel or cause for alarm. The climate is always cooling or warming, becoming wetter or drier. The problem with the current cycle of change is its rate and magnitude, which appear unusual in the context of the last several millennia. So what's causing this shift? Although scientists recognize the importance of nonanthropogenic (i.e., "natural") drivers of change, there is consensus that **anthropogenic** (i.e., human) activities have been a dominant force in recent climate change (IPCC 2007c).

Drivers of climate change are typically categorized as forcings or feedbacks. **Radiative forcing** is a measure of positive (warming) or negative (cooling) changes to Earth's energy balance. **Forcing agents** (e.g., greenhouse gases and volcanic dust) affect the energy balance by altering the amount of incoming solar radiation, reflected solar radiation, or outgoing longwave radiation. **Feedbacks** are internal processes in the climate system that either amplify or diminish the effects of a forcing agent (Le Treut et al. 2007).

Figure 2.5

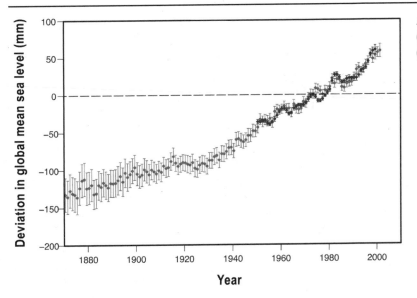

Annual deviation of global sea level (mm) from the long-term mean (average of all years from 1961–1990).

Years with higher-than-average sea level appear above the dotted reference line, whereas years with lower-than-average sea level appear below the line. The vertical bars above and below each data point represent the 90% confidence interval (i.e., scientists have 90% confidence that the true value of the sea level lies somewhere between the maximum and minimum points on the bars). Before 1950, there were fewer than 100 tide gauges around the world. Therefore, the light gray curve shows sea levels reconstructed from tide gauges and models that include modern-day satellite data. The dark gray curve shows coastal tide gauge measurements since 1950. The close match between the reconstructed and observed sea levels gives scientists a high level of confidence that the models are representative of years prior to 1950. See NOS Box 2.4 for an explanation of how (and why) deviations are calculated.

Source: Bindoff et al. 2007. Modified with permission.

Forcings
CHANGES IN EARTH'S ORBIT

The geometric relationship between our planet and the Sun is constantly changing: The seasonal direction of Earth's axis varies on 19,000-year and 23,000-year cycles, the angle of Earth's tilt varies on a 41,000-year cycle, and the shape of Earth's orbit becomes more elliptical or more circular on a 100,000-year cycle. These cycles are named the **Milankovitch cycles**, after the Serbian engineer and mathematician who calculated changes in Earth's orbit in the 1930s. The cycles affect the amount of solar radiation received in each hemisphere, and they appear to trigger ice ages when solar radiation in high northern latitudes drops below a critical threshold (Figure 2.2). The Milankovitch cycles are not responsible for the planet's current warming trend. First, changes in Earth's orbit affect climate very slowly, over thousands to tens of thousands of years. Second, we are entering the cooling phase of the cycles; scientists expect the next low in solar radiation in about 30,000 years (Jansen et al. 2007).

SOLAR ACTIVITY

The global radiation budget fluctuates because of an 11-year cycle in sunspot activity. **Sunspots** are relatively cool, dark areas on the surface of the Sun that last days to weeks. They are accompanied by hotter-than-average regions called **faculae** that increase the overall radiative output of the Sun (Jansen et al. 2007). Since the Maunder Minimum, a period from about AD 1645 to 1715 when there was negligible sunspot activity, solar irradiance has increased by 0.08%. The positive forcing (warming) effect of solar radiation seems to be about 10 times less than forcing due to greenhouse gas emissions (Figure 2.8, p. 38) (Forster et al. 2007). If sunspots *were* a dominant contributor to climate change, scientists would expect increases in daily maximum temperatures but little or no change in nighttime minimum temperatures. Instead, because of an enhanced greenhouse effect—which acts both day and night—scientists have observed similar significant upward trends in both maximum and minimum temperatures (Wild, Ohmura, and Makowski 2007).

AEROSOLS

Aerosols are small particles that are either directly emitted into the atmosphere or formed by the chemical conversion of emitted gases. Although many nonanthropogenic processes release aerosols, human activities such as mining and agriculture have increased the amount of dust, sulfates, organic compounds, and soot in the atmosphere.

Depending on their type, aerosols may exert a warming effect by absorbing solar and longwave radiation or a cooling effect by reflecting radiation or increasing the albedo of clouds. For example, volcanic aerosols tend to cause regional or global cooling. When Mt. Pinatubo erupted in 1991 (Figure 2.6a), it projected sulfate aerosols, ash, and dust above the highest clouds (Figure 2.6b). The aerosols traveled around the world, reducing average global temperature by about 0.5°C. It appears that all anthropogenic and nonanthropogenic aerosols combined have a small net cooling effect (Figure 2.8) (Forster et al. 2007).

GREENHOUSE GASES

Long-lived greenhouse gases, which persist in the atmosphere for years to decades, are the main cause of Earth's current warming trend

Figure 2.6

(a) Eruption of Mt. Pinatubo

(b) Before and after eruption

September 1984

August 1991

(a) Eruption of Mt. Pinatubo, Philippines, on June 12, 1991. Photograph was taken about 25 km east of the volcano.

(b) Space shuttle images of the atmosphere before and after the eruption. The atmosphere appears thicker in the September 1991 image because of the addition of ash and sulfate aerosols.

Photo (a) courtesy of USGS/Cascades Volcano Observatory.

Photos (b) courtesy of NASA—Goddard Space Flight Center Scientific Visualization Studio.

(Figure 2.7, p. 38) (Forster et al. 2007). In this section we will consider only carbon dioxide (CO_2) and methane (CH_4), although several other gases such as nitrous oxide, a host of halocarbon compounds, and ozone also enhance the natural greenhouse effect. Water vapor, a **short-lived greenhouse gas**, can cause either regional warming or cooling through feedback loops, and will be considered later.

Carbon dioxide can remain in the atmosphere from 5 to 200 years and has more effect on Earth's temperature than any other long-lived greenhouse gas (Pitman 2005). From AD 1000 to 1750, atmospheric CO_2 fluctuated between 275 and 285 **parts per million** (ppm). (One ppm of CO_2 is just over 2 Gt of carbon.) By 2005, there were 379 ppm of CO_2 in the atmosphere, a 36% increase over 250 years. In just the last decade, we have seen the highest average increase in CO_2 since 1950 (Figure 2.7, p. 36) (Forster et al. 2007). About 80% of the CO_2 increase has come from fossil fuel use and cement production. The remainder has come from changes in land use, especially tropical deforestation (Denman et al. 2007).

Carbon is moving constantly between the land, ocean, atmosphere, and living things. Scientists can identify the source of the extra carbon in the atmosphere with the **carbon isotope ratio.** There are two naturally occurring, stable isotopes of carbon referred to as ^{13}C

Figure 2.7

Trend in atmospheric concentration of carbon dioxide from the last two thousand years.

Source: Forster et al. 2007. Modified with permission.

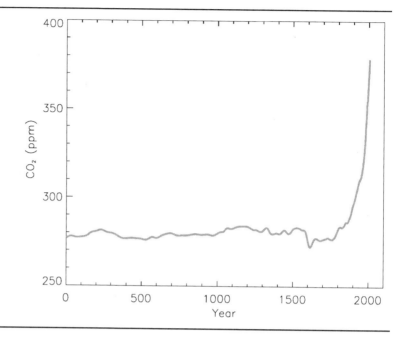

and ^{12}C. Photosynthesis uses both, but takes in more ^{12}C than would be expected based on the composition of the atmosphere. Therefore, the CO_2 released from decomposing or burning plant matter (like coal, which is made of fossilized plants) has a higher proportion of ^{12}C (i.e., a lower carbon isotope ratio) than the CO_2 in the atmosphere, oceans, or geothermal emissions. The carbon isotope ratio has been declining in the atmosphere since the late 1970s (Forster et al. 2007).

About 55% of the CO_2 released by human activities since 1959 has been taken up by **carbon sinks**, areas of the Earth that temporarily remove CO_2 from the atmosphere. The biggest sink is the ocean, which has taken up about 30% of the CO_2. Dissolved CO_2 stays in surface waters for fewer than 10 years because it is transported to the bottom by cold water and sinking particles of dead phytoplankton. The remaining 25% of the CO_2 has been taken up by terrestrial plants and soils (i.e., through photosynthesis) (Denman et al. 2007). Some of the most important terrestrial sinks are conifer forests in northwestern Canada, tropical rainforests in South America and Africa, and boreal forests in the Northern Hemisphere (CDIAC 2000).

Methane, which stays in the atmosphere for 9 to 15 years, is the second most important long-lived greenhouse gas. Although there are many nonhuman sources of CH_4, such as volcanoes and wetlands, human activities such as agriculture and landfills are

36

responsible for large increases in this gas. In the last 250 years, atmospheric CH_4 rose from about 700 parts per billion (ppb) to 1,775 ppb. The rate of CH_4 increase was five times faster from 1960 to 1999 than any 40-year period between AD 1600 and 1800 (Jansen et al. 2007).

Ice core data suggest the current atmospheric concentration of CO_2 and CH_4 are higher than any other time in the last 650,000 years (Figure 2.2). Furthermore, the rate of increase has been higher than any other time in at least the last 16,000 years (Jansen et al. 2007).

LAND COVER

Changes in land use strongly influence local surface climates through shifts in radiation balance, stored heat, and cloudiness. In cities, for example, humidity falls and temperature rises when roads and parking lots replace vegetated surfaces (which is one aspect of the urban heat island effect; see Chapters 1 and 10). The reverse happens when irrigated vegetation such as golf courses replace bare surfaces (Denman et al. 2007).

Topic: Explore carbon
Go to: www.scilinks.org
Code: CCPP11

Overall, changes in land use have probably had a slight cooling effect on the climate by increasing global albedo. (The release of carbon by deforestation and agriculture, which has had a warming effect, is considered separately; see "Greenhouse Gases," p. 34.) In the 20th century, humans reduced forest cover by about 11 million km^2 and increased cropland by 38–42 million km^2. Forest generally has a lower albedo than open land due to its dark color, greater leaf area, and higher structural complexity. Forest also tends to remain above snow cover in the winter, whereas crops and other short vegetation are covered by snow (Forster et al. 2007).

SUMMARY OF FORCINGS

The global-scale warming trend of the first half of the 20th century seems to be explained in roughly equal measure by a decrease in volcanic activity, changes in internal climate variability (e.g., El Niño), and increased greenhouse gas emissions. Changes in the solar cycle also appear to have contributed a marginal amount of warming (Hegerl et al. 2007a). The warming of the past 50 years, however, has been dominated by anthropogenic effects; long-lived greenhouse gases account for about 75% of the global trend (Figure 2.8, p. 38) (Forster et al. 2007). To achieve greater confidence in this result, scientists have simulated the temperature trend of the past 100 years with two models: one that included only nonanthropogenic forcings and another with both nonanthropogenic and anthropogenic forcings (Figure 2.9, p. 38). The "natural" forcings model failed to fit the observed warming trend, and in fact suggested Earth should have

Figure 2.8

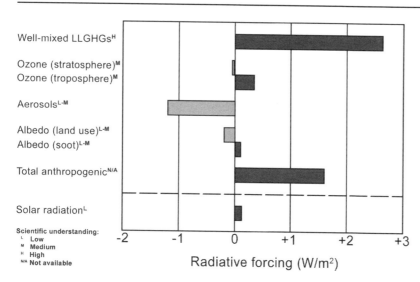

Relative rate of energy change (watts) per unit area (square meter) of the globe caused by anthropogenic and solar forcing agents since the start of the industrial era (AD 1750).

Positive forcings warm the climate; negative forcings cool the climate. Only forcings >0.05 W/m² are shown. The aerosol bar includes both direct and indirect effects (i.e., changes to clouds). The level of scientific understanding for each forcing, indicated with superscripts on the y-axis, is based on the strength of evidence and the amount of scientific consensus.

Source: Forster et al. 2007. Modified with permission.

Figure 2.9

Relative effects of nonhuman forcings (e.g., natural aerosols) vs. human forcings (e.g., fossil fuel emissions) plus nonhuman forcings.

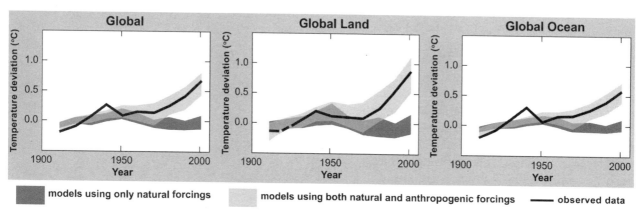

The black lines represent observed temperature changes (i.e., from the instrumental record) relative to 1961–1990, and the gray bands represent model simulations of observed climate.

Source: Hegerl et al. 2007b. Modified with permission.

cooled slightly. The model that included human effects, on the other hand, described the data from the instrumental record relatively well (Hegerl et al. 2007b).

In a similar analysis, scientists from the United Kingdom attributed global increases in **specific humidity** (mass of water vapor in the atmosphere) to a mix of anthropogenic and nonanthropogenic forcings. Models that contained both types of forcings closely matched the observed upward trend in global specific humidity over the last 30 years (Willett et al. 2007).

Feedbacks

Feedbacks are cyclic processes that greatly contribute to the rate and magnitude of regional climate change. Unfortunately, scientists have difficulty modeling regional change because of seasonal and spatial variation in feedbacks (Denman et al. 2007). In the Arctic, for example, scientists have been surprised by the rapid loss of sea ice. Although climate models project a decline in ice, they have thus far underestimated the observed rate of loss (M. Serreze, Interview on National Public Radio, September 21, 2007). Below, we discuss some of the climate feedbacks under intense investigation.

VEGETATION

In seasonal environments, more CO_2 and higher temperatures often cause an earlier onset of plant growth, a longer growing season, and higher productivity of plants. These changes typically result in a positive feedback. Take, for example, the Arctic, where the snow season has been shortened and plant productivity has increased. As vegetation cover increases, rates of evapotranspiration increase. As the amount of water vapor increases, local rainfall increases. As it gets warmer and wetter, more vegetation grows, and the feedback cycle continues (Zhang and Walsh 2006).

Changes in vegetation also can amplify warming if they increase the frequency and intensity of wildfires. For instance, cheatgrass (*Bromus tectorum*), an annual grass that has a strong positive response to elevated CO_2, is invading shrublands in the western United States. As cheatgrass spreads, it creates a higher fuel load and increases the risk of fire. As more vegetation is burned, more CO_2 is released and warming is enhanced (Bradley et al. 2006).

Scientists are fairly certain that terrestrial vegetation will, overall, warm the climate because of processes such as the ones detailed above. However, there are many unknowns, including the long-term effects of CO_2 enrichment on plants and how carbon storage might change in soils (Denman et al. 2007).

CRYOSPHERE

The melting of the cryosphere creates three positive feedback loops. First, land and water have a lower albedo than snow and ice, so they tend to absorb more shortwave radiation and then emit more longwave radiation back to the atmosphere. Second, evaporation of melt water increases local concentrations of water vapor, which is a greenhouse gas. Third, carbon in organic compounds is released to the atmosphere as permafrost thaws (Lemke et al. 2007).

However, scientists are not yet sure of the effects of melting sea ice on cloud formation (due to increased evaporation and aerosols; see below) or on global circulation. Recall from Chapter 1 that movements of the atmosphere and ocean are due in large part to the energy gradient between the poles and the equator. It is uncertain how the weakening of that gradient might affect global climate (Randall et al. 2007).

WATER VAPOR AND CLOUDS

As the climate warms, the concentration of water vapor in the atmosphere increases because evapotranspiration happens more rapidly and because warm air can hold more water (about 7% more for every 1°C increase in temperature) (Trenberth et al. 2007). In the last 30 years, the global mean specific water vapor has increased by 0.07 g/kg (Willett et al. 2007). Because water vapor is such an important greenhouse gas, it creates a large positive feedback (Randall et al. 2007).

However, a moister atmosphere also can be cloudier, and clouds can cause local warming *or* cooling. Whether a cloud absorbs longwave radiation or reflects shortwave radiation depends on its location, altitude, aerosol concentration, and many other factors (Le Treut et al. 2007). Currently, it seems the overall effect of clouds is a small negative feedback. It is unknown how long this cooling effect will last, if at all (Randall et al. 2007).

WHY IS CLIMATE CHANGE IMPORTANT?

As we discussed in Chapter 1, climatic conditions—and therefore changes in climate—can impact the health and reproductive success of all organisms. Although the complexity of biological systems makes it difficult to study the effects of climate change, scientists already have observed structural and functional shifts in ecosystems worldwide. In Chapter 3, we will set the stage for the classroom activities in Part II of the book by summarizing how climate change has affected biological systems thus far.

REFERENCES

American Association for the Advancement of Science (AAAS). 2007. AAAS Board statement on climate change. Retrieved October 3, 2007, from *www.aaas.org/news/press_room/climate_change/mtg_200702/aaas_climate_statement.pdf*

American Geophysical Union (AGU). 2008. Human impacts on climate. Retrieved September 6, 2008, from *www.agu.org/sci_soc/policy/positions/climate_change2008.shtml*

American Meteorological Society (AMS). 2000. *Glossary of meteorology.* Electronic version of the second edition. Retrieved October 3, 2007, from *http://amsglossary.allenpress.com/glossary*

Atlantic Oceanographic and Meteorological Laboratory (AOML). 2007. Hurricane Research Division: Frequently asked questions. Retrieved November 21, 2007, from *www.aoml.noaa.gov/hrd/tcfaq/tcfaqHED.html*

Bindoff, N. L., J. Willebrand, V. Artale, A. Cazenave, J. Gregory, S. Gulev, K. Hanawa, C. Le Quéré, S. Levitus, Y. Nojiri, C. K. Shum, L. D. Talley, and A. Unnikrishnan. 2007. Observations: Oceanic climate change and sea level. In *Climate change 2007: The physical science basis. Contribution of Working Group I to the Fourth Assessment Report of the Intergovernmental Panel on Climate Change,* eds. S. Solomon, D. Qin, M. Manning, Z. Chen, M. Marquis, K. B. Averyt, M. Tignor, and H. L. Miller, 385–432. Cambridge: Cambridge University Press.

Bowen, M. 2005. *Thin ice: Unlocking the secrets of climate in the world's highest mountains.* New York: Henry Holt.

Bradley, B. A., R. A. Houghton, J. F. Mustard, and S. P. Hamburg. 2006. Invasive grass reduces aboveground carbon stocks in shrublands of the western U.S. *Global Change Biology* 12(10): 1815–1822.

Carbon Dioxide Information Analysis Center (CDIAC). 2000. Major world ecosystem complexes ranked by carbon in live vegetation. Retrieved November 21, 2007, from *http://cdiac.ornl.gov/ftp/ndp017/table.html*

Dalmasso, P. R., R. A. Taccone, J. D. Nieto, M. A. Teruel, and S. I. Lane. 2006. CH_3OCF_2CHFCl and CHF_2OCF_2CHFCl: Reaction with Cl atoms, atmospheric lifetimes, ozone depletion and global warming potentials. *Atmospheric Environment* 40(38): 7298–7307.

Denman, K. L., G. Brasseur, A. Chidthaisong, P. Ciais, P. M. Cox, R. E. Dickinson, D. Hauglustaine, C. Heinze, E. Holland, D. Jacob, U. Lohmann, S. Ramachandran, P. L. da Silva Dias, S. C. Wofsy, and X. Zhang. 2007. Couplings between changes in the climate system and biogeochemistry. In *Climate change*

2007: The physical science basis. Contribution of Working Group I to the Fourth Assessment Report of the Intergovernmental Panel on Climate Change, eds. S. Solomon, D. Qin, M. Manning, Z. Chen, M. Marquis, K. B. Averyt, M. Tignor, and H. L. Miller, 499–587. Cambridge: Cambridge University Press.

Emanuel, K. A. 1987. The dependence of hurricane intensity on climate. *Nature* 326 (6112): 483–485.

Emanuel, K., R. Sundararajan, and J. Williams. 2008. Hurricanes and global warming: Results from downscaling IPCC AR4 simulations. *Bulletin of the American Meteorological Society* 89(3): 347–367.

Forster, P., V. Ramaswamy, P. Artaxo, T. Berntsen, R. Betts, D. W. Fahey, J. Haywood, J. Lean, D. C. Lowe, G. Myhre, J. Nganga, R. Prinn, G. Raga, M. Schulz, and R. Van Dorland. 2007. Changes in atmospheric constituents and in radiative forcing. In *Climate change 2007: The physical science basis. Contribution of Working Group I to the Fourth Assessment Report of the Intergovernmental Panel on Climate Change*, eds. S. Solomon, D. Qin, M. Manning, Z. Chen, M. Marquis, K. B. Averyt, M. Tignor, and H. L. Miller, 129–234. Cambridge: Cambridge University Press.

Gauch, H. G. 2003. *Scientific method in practice*. Cambridge: Cambridge University Press.

Hegerl, G. C., T. J. Crowley, M. Allen, W. T. Hyde, H. N. Pollack, J. Smerdon, and E. Zorita. 2007a. Detection of human influence on a new, validated 1500-year temperature reconstruction. *Journal of Climate* 20(4): 650–666.

Hegerl, G. C., F. W. Zwiers, P. Braconnot, N. P. Gillett, Y. Luo, J. A. Marengo Orsini, N. Nicholls, J. E. Penner, and P. A. Stott. 2007b. Understanding and attributing climate change. In *Climate change 2007: The physical science basis. Contribution of Working Group I to the Fourth Assessment Report of the Intergovernmental Panel on Climate Change*, eds. S. Solomon, D. Qin, M. Manning, Z. Chen, M. Marquis, K. B. Averyt, M. Tignor, and H. L. Miller, 663–745. Cambridge: Cambridge University Press.

Hinzman, L. D., N. D. Bettez, W. R. Bolton, F. S. Chapin, M. B. Dyurgerov, C. L. Fastie, B. Griffith, R. D. Hollister, A. Hope, H. P. Huntington, A. M. Jensen, G. J. Jia, T. Jorgenson, D. L. Kane, D. R. Klein, G. Kofinas, A. H. Lynch, A. H. Lloyd, A. D. McGuire, F. E. Nelson, W. C. Oechel, T. E. Osterkamp, C. H. Racine, V. E. Romanovsky, R. S. Stone, D. A. Stow, M. Sturm, C. E. Tweedie, G. L. Vourlitis, M. D. Walker, D. A. Walker, P. J. Webber, J. M. Welker, K. S. Winker, and K. Yoshikawa. 2005. Evidence and implications of recent climate change in northern Alaska and other Arctic regions. *Climatic Change* 72(3): 251–298.

Hoegh-Guldberg, O., P. J. Mumby, A. J. Hooten, R. S. Steneck, P. Greenfield, E. Gomez, C. D. Harvell, P. F. Sale, A. J. Edwards, K. Caldeira, N. Knowlton, C. M. Eakin, R. Iglesias-Prieto, N. Muthiga, R. H. Bradbury, A. Dubi, M. E. Hatziolos. 2007. Coral reefs under rapid climate change and ocean acidification. *Science* 318 (5857): 1737–1742.

Hobgood, J. 2005. Tropical cyclones. In *Encyclopedia of world climatology*, ed. J. E. Oliver, 750–756. Dordrecht, The Netherlands: Springer.

Intergovernmental Panel on Climate Change (IPCC). 2007a. *Climate change 2007: Synthesis report. Contribution of Working Groups I, II and III to the Fourth Assessment Report of the Intergovernmental Panel on Climate Change*, eds. Core Writing Team, R. K. Pachauri and A. Reisinger. Geneva, Switzerland: IPCC.

Intergovernmental Panel on Climate Change (IPCC). 2007b. Fact sheet. Retrieved November 21, 2007 from *www.ipcc.ch/press/ ar4-factsheet1.htm*

Intergovernmental Panel on Climate Change (IPCC). 2007c. Summary for policymakers. In *Climate change 2007: The physical science basis. Contribution of Working Group I to the Fourth Assessment Report of the Intergovernmental Panel on Climate Change*, eds. S. Solomon, D. Qin, M. Manning, Z. Chen, M. Marquis, K. B. Averyt, M. Tignor, and H. L. Miller, 1–18. Cambridge: Cambridge University Press.

Jansen, E., J. Overpeck, K. R. Briffa, J.-C. Duplessy, F. Joos, V. Masson-Delmotte, D. Olago, B. Otto-Bliesner, W. R. Peltier, S. Rahmstorf, R. Ramesh, D. Raynaud, D. Rind, O. Solomina, R. Villalba, and D. Zhang. 2007. Palaeoclimate. In *Climate change 2007: The physical science basis. Contribution of Working Group I to the Fourth Assessment Report of the Intergovernmental Panel on Climate Change*, eds. S. Solomon, D. Qin, M. Manning, Z. Chen, M. Marquis, K. B. Averyt, M. Tignor, and H. L. Miller, 433–497. Cambridge: Cambridge University Press.

Joint Science Academies. 2005. Global response to climate change. Retrieved October 3, 2007, from *http://nationalacademies.org/ onpi/06072005.pdf*

Jones, P. D., and M. E. Mann. 2004. Climate over past millennia. *Reviews of Geophysics* 42(2): RG2002, doi:10.1029/2003RG000143.

Jones, P. D., M. New, D. E. Parker, S. Martin, and I. G. Rigor. 1999. Surface air temperature and its changes over the past 150 years. *Reviews of Geophysics* 37(2): 173–199.

Le Treut, H., R. Somerville, U. Cubasch, Y. Ding, C. Mauritzen, A. Mokssit, T. Peterson, and M. Prather. 2007. Historical overview of climate change. In *Climate change 2007: The physical science basis.*

Contribution of Working Group I to the Fourth Assessment Report of the Intergovernmental Panel on Climate Change, eds. S. Solomon, D. Qin, M. Manning, Z. Chen, M. Marquis, K. B. Averyt, M. Tignor, and H. L. Miller, 93–127. Cambridge: Cambridge University Press.

Lemke, P., J. Ren, R. B. Alley, I. Allison, J. Carrasco, G. Flato, Y. Fujii, G. Kaser, P. Mote, R. H. Thomas, and T. Zhang. 2007. Observations: Changes in snow, ice and frozen ground. In *Climate change 2007: The physical science basis. Contribution of Working Group I to the Fourth Assessment Report of the Intergovernmental Panel on Climate Change*, eds. S. Solomon, D. Qin, M. Manning, Z. Chen, M. Marquis, K. B. Averyt, M. Tignor, and H. L. Miller, 337–383. Cambridge: Cambridge University Press.

McCarthy, L. 2008. 2007 was tied as Earth's second-warmest year. Retrieved May 3, 2008 from *www.nasa.gov/centers/goddard/ news/topstory/2008/earth_temp.html*

McPhaden, M. J., S. E. Zebiak, and M. H. Glantz. 2006. ENSO as an integrating concept in earth science. *Science* 314(5806): 1740–1745.

Meier, M. F., M. B. Dyurgerov, U. K. Rick, S. O'Neel, W. T. Pfeffer, R. S. Anderson, S. P. Anderson, and A. F. Glazovsky. 2007. Glaciers dominate eustatic sea-level rise in the 21st century. *Science* 317(5841): 1064–1067.

National Snow and Ice Data Center (NSIDC). 2007. Arctic sea ice shatters all previous record lows. Retrieved November 21, 2007, from *http://nsidc.org/news/press/2007_ seaiceminimum/20071001_pressrelease.html*

Palassis, J. 1999. Portable infrared analyzers. *Applied Occupational and Environmental Hygiene* 14(8): 510–514.

Pielke, R. A. 2005. Consensus about climate change? *Science* 308(5724): 952–953.

Pitman, A. J. 2005. Greenhouse effect and greenhouse gases. In *Encyclopedia of world climatology*, ed. J. E. Oliver, 391–397. Dordrecht, The Netherlands: Springer.

Randall, D. A., R. A. Wood, S. Bony, R. Colman, T. Fichefet, J. Fyfe, V. Kattsov, A. Pitman, J. Shukla, J. Srinivasan, R. J. Stouffer, A. Sumi, and K. E. Taylor. 2007. Climate models and their evaluation. In *Climate change 2007: The physical science basis. Contribution of Working Group I to the Fourth Assessment Report of the Intergovernmental Panel on Climate Change*, eds. S. Solomon, D. Qin, M. Manning, Z. Chen, M. Marquis, K. B. Averyt, M. Tignor, and H. L. Miller, 589–662. Cambridge: Cambridge University Press.

Rosenzweig, C., G. Casassa, D. J. Karoly, A. Imeson, C. Liu, A. Menzel, S. Rawlins, T. L. Root, B. Seguin, and P. Tryjanowski.

2007. Assessment of observed changes and responses in natural and managed systems. In *Climate change 2007: Impacts, adaptation and vulnerability. Contribution of Working Group II to the Fourth Assessment Report of the Intergovernmental Panel on Climate Change*, eds. M. L. Parry, O. F. Canziani, J. P. Palutikof, P. J. van der Linden, and C. E. Hanson, 79–131. Cambridge: Cambridge University Press.

Saltzman, B. 2002. *Dynamical paleoclimatology: Generalized theory of global climate change*. San Diego: Academic Press.

Serreze, M. C., M. M. Holland, and J. Stroeve. 2007. Perspectives on the Arctic's shrinking sea-ice cover. *Science* 315(5818): 1533–1536.

Thoning, K. W., D. R. Kitzis, and A. Crotwell. 2007. Atmospheric carbon dioxide dry air mole fractions from quasi-continuous measurements at Barrow, Alaska; Mauna Loa, Hawaii; American Samoa; and South Pole, 1973–2006. Version: 2007-10-01. Retrieved November 21, 2007, from *ftp://ftp.cmdl.noaa.gov/ccg/co2/in-situ*

Trenberth, K. E., P. D. Jones, P. Ambenje, R. Bojariu, D. Easterling, A. Klein Tank, D. Parker, F. Rahimzadeh, J. A. Renwick, M. Rusticucci, B. Soden, and P. Zhai. 2007. Observations: Surface and atmospheric climate change. In *Climate change 2007: The physical science basis. Contribution of Working Group I to the Fourth Assessment Report of the Intergovernmental Panel on Climate Change*, eds. S. Solomon, D. Qin, M. Manning, Z. Chen, M. Marquis, K. B. Averyt, M. Tignor, and H. L. Miller, 235–336. Cambridge: Cambridge University Press.

Weart, S. R. 2003. *The discovery of global warming*. Cambridge, MA: Harvard University Press. Also available at: *http://aip.org/history/climate*

Wild, M., A. Ohmura, and K. Makowski. 2007. Impact of global dimming and brightening on global warming. *Geophysical Research Letters* 34(4): L04702, doi:10.1029/2006GL028031.

Willett, K. M., N. P. Gillett, P. D. Jones, and P. W. Thorne. 2007. Attribution of observed surface humidity changes to human influence. *Nature* 449(7163): 710–713.

Zhang, J., and J. E. Walsh. 2006. Thermodynamic and hydrological impacts of increasing greenness in northern high latitudes. *Journal of Hydrometeorology* 7(5): 1147–1163.

Chapter 3
Biological Effects of Climate Change

"Many of the new climates will include combinations of temperature, precipitation, seasonality, and day length that do not currently exist anywhere on Earth.... Something will live in these nonanalogue climates, but it is difficult to guess what."

—Chris D. Thomas, "Recent Evolutionary Effects of Climate Change" in *Climate Change and Biodiversity*, 2005

Species on Earth face many threats. Their habitats are destroyed by agriculture, logging, and mining. Their physical environments are poisoned by heavy metals and pesticides. They are harvested for food or fiber, or are collateral damage when other species are harvested. Against this backdrop, how important is climate change—something that has occurred throughout Earth's history? Can ecosystems tolerate the magnitude and rate of future change? How will other conservation threats interact with climate change? How likely are widespread extinctions, and how might they affect the functioning of ecosystems and human societies? Biologists are trying to answer these and other questions as climate change intensifies.

Despite the unknowns, it is clear that human-dominated climate change has already left its fingerprint on biological systems. Using a combination of new observations and datasets spanning decades to centuries, scientists have detected responses across the globe and in most major taxonomic groups (Parmesan 2006).

HOW ARE BIOLOGICAL EFFECTS DETECTED?

Because global ecosystems face so many threats, it can be difficult to attribute changes in a biological system to rising temperatures or changing precipitation. Scientists must demonstrate a convincing relationship between biotic responses and regional, rather than global, climate trends (Rosenzweig et al. 2007). Like climatologists, biologists cannot do manipulative studies on entire ecosystems or biomes (Chapter 2). Instead, they rely on a combination of observations, small-scale experiments, and simulation models.

Observational Studies

The traditional (and perhaps most common) approach to studying the effects of climate change is to make long-term observations (i.e., more than 10 years) of systems that have not been experimentally manipulated. Observations may be continuous, as in a naturalist's diary, or may consist of multiple surveys of the same place (Thuiller 2007). Scientific observations also can be complemented by **traditional ecological knowledge** (NOS Box 3.1). In addition to trends over time and space, researchers are interested in how biological systems react to specific climate elements, such as temperature. One way to address this question is to examine biological systems over **environmental gradients** (e.g., warm to cold, dry to wet, etc.) (Dunne et al. 2004).

In paleoecological studies, researchers use the fossil record to examine biological responses to past climates. Because most species of the last 2–3 million years are the same or similar to modern species, paleoecology can suggest the likely speed and mechanism of responses to modern climate change (Jackson and Williams 2004).

A correlation between a biological response and an anthropogenic climate change does not automatically imply a causal relation-

Nature of Science (NOS) Box 3.1. Scientific knowledge can be complemented by traditional knowledge.

To gain an in-depth understanding of a natural system, a scientist generally needs to study that system for 10 years or more. Unfortunately, there are numerous constraints on long-term studies. Lack of funding, changes in institutional interest, pressure to publish results, or even the death of a researcher can limit opportunities to collect data over multiple years.

In some ecosystems, researchers can alleviate this problem by tapping into traditional ecological knowledge (TEK). TEK consists of decades to hundreds of years of observations made by native peoples living in close association with their natural environment. Although the observations tend to be qualitative and local in scale, they span more time than the average scientific study. Scientists can use TEK to gauge confidence in their results, identify new research questions, and examine potential mechanisms that explain both sets of observations (Huntington et al. 2004). For example, some scientists studying sea ice in the Arctic are trying to use Inuit knowledge to evaluate and improve regional climate models (Laidler 2006).

ship. Nonclimatic factors or natural climate variability actually may be causing the response. For example, both the El Niño-Southern Oscillation (ENSO; see Chapter 1), which is a natural climate cycle, and anthropogenic climate change can alter the structure of plant communities or the geographic range of marine fish. Scientists need decades of data and advanced statistical techniques to separate the effects of these two phenomena (Rosenzweig et al. 2007). Scientists also can increase confidence in a causal relationship by performing experiments (see next section), studying multiple species at once (Parmesan and Yohe 2003), and confirming that relationships are consistent over time and space (Kerr, Kharouba, and Currie 2007). You can use the activities in Chapter 8 to explore correlation versus causation with your students.

Experimental Studies

Scientists use controlled experiments to identify causes of biological responses to climate change and to inform projections of future responses. Experiments can be carried out in the field or in the laboratory (Dunne et al. 2004). For example, researchers can manipulate carbon dioxide (CO_2) concentrations in small growth chambers within a laboratory or in greenhouses. Scientists also can manipulate CO_2 in the field with large Free Air CO_2 Enrichment (FACE) systems (Denman et al. 2007). In a FACE system, vertical pipes blow CO_2 into a circular outdoor plot (Figure 3.1, p. 50). Unlike in a lab or a greenhouse, plants in a FACE system are still exposed to sunlight, wind, rain, herbivores, and pests (BNL 2006).

Despite its importance, experimentation has a number of drawbacks. First, experiments can be expensive and time-consuming. It

Figure 3.1

Free Air CO$_2$ Enrichment system in the Blackwood Division of Duke Forest, near Chapel Hill, North Carolina.

The ecosystem being studied is a loblolly pine (*Pinus taeda*) forest plantation.

Photograph © Duke Photography. Reprinted with permission.

also is difficult to balance realism (i.e., by including multiple variables) and ease of interpretation. Finally, it's hard to use the experimental results from one system or scale to infer change in another, because of historical factors, regional climate processes, and other issues. Scientists can minimize these problems by combining multiple techniques (NOS Box 3.2).

Modeling Studies

Scientists use simulation models to forecast the future biological effects of climate change. One type of model forecasts future effects based on abiotic factors alone. For instance, the gypsy moth (*Lymantria dispar*) is a Eurasian pest that is introduced occasionally to the western United States but has not yet established self-sustaining populations. To successfully complete its life cycle, the moth needs cool winters and warm summers. Temperature-based simulation models suggest that the warming climate is increasing the risk that gypsy moths will become established in Utah (Logan et al. 2007). A

Nature of Science (NOS) Box 3.2. There is no one "scientific method."

Textbooks often describe science as a linear process that moves from a clear hypothesis, to experimental manipulation, to a set of tidy results. This simplified description, however, fails to convey the true nature of science. For one thing, the development of a hypothesis is sometimes difficult. It can require intuition, creativity, and even trial and error (Singer 2007). Second, the scientific process is more like a loop than a line. Researchers use both observations to generate principles and general principles to explain observations (Barnosky and Kraatz 2007). Finally, the results of a study can be difficult to interpret because natural systems and phenomena are so complex.

The methods used in a scientific study depend on the questions being asked, logistical and/or ethical constraints, and the interests and expertise of the scientist. Often, researchers will use multiple techniques to get at different aspects of the same question. For example, a researcher in Washington used the following methods to study climate-induced range expansion of a butterfly called the sachem skipper (*Atalopedes campestris*) (Crozier 2002):

- Use of historical collection records to reconstruct the butterfly's western range;
- Field surveys to determine the current expansion of the butterfly's range;
- Behavioral observations of caterpillars;
- Transplant experiments, in which butterflies were relocated outside their current range;
- Laboratory tests of the butterfly's tolerance to cold; and
- Comparison of the butterfly's macro- and microclimates.

more complicated type of model forecasts effects based on interactions within and between species and the availability of living and nonliving resources (Thuiller 2007). For example, biologists in Florida forecast the future range of a pest species called "red imported fire ants" (*Solenopsis invicta*) with a model that combined soil temperature, colony size, and the dispersal ability of individuals within each colony (Morrison, Korzukhin, and Porter 2005). Their model suggests that the habitat of these ants will increase by 5% in the next 40–50 years and more than 21% in the next 100 years.

HOW HAVE BIOLOGICAL SYSTEMS CHANGED?

As we discussed in Chapter 1, organisms can survive, grow, and reproduce in only a limited range of environmental conditions (Figure 1.4, p. 11). If changes in climate exceed the natural variability of a species' preferred environment, there can be direct and indirect effects from individuals to ecosystems. Because of differences in physiology, dispersal ability, population size, and other factors, no two species will respond identically to climate change (Parmesan 2006). Table 3.1, page 52, lists examples of some of the responses to climate change that researchers have documented already.

Table 3.1

Examples of observed responses in wild biological systems to climate change. Responses may be due to underlying genetic change or flexibility in physiology, behavior, or life history. Source: Compiled from McCarty 2001; Parmesan 2006; Rosenzweig et al. 2007.	**Physiology and morphology**	• Larger body size • Higher metabolic demand • Shorter life cycles • Heat stress/mortality • Drought stress/mortality
	Phenology	• Earlier leaf unfolding and flowering • Earlier fruit ripening • Earlier/later arrival of migratory species • Earlier reproductive activities • Earlier emergence from dormancy/hibernation • Mismatch between life cycles of organisms and their pollinators or food sources
	Abundance and distribution	• Population declines and local extinctions • Population increases • Range expansions • Range shifts • Range contractions
	Ecosystem structure and function	• Loss/replacement of native species • Invasion by non-native species • Decreases/increases in plant productivity • Accelerated carbon cycling

Physiology and Morphology

The physiological effects of climate change depend on the severity and rate of the change and the vulnerability of organisms to that change. When a population is exposed to physiological stress, it may **acclimatize** and become more tolerant of the stress in days or weeks, undergo genetic change over many generations, or become locally extinct (Helmuth, Kingsolver, and Carrington 2005).

The effects of a warming climate depend in part on the sensitivity of critical biological functions to temperature. For example, most sea turtles (e.g., green turtles, *Chelonia mydas*; Figure 3.2) undergo **temperature dependent sex determination**, in which incubation temperature determines the gender of hatchlings. Above a certain temperature (29°C in green turtles), sea turtle nests produce mostly females; below that temperature, nests produce mostly males. On Ascension Island, in the South Atlantic, a range of incubation temperatures are available to turtles because of differences in beach color and, hence, albedo (see Chapter 1, p. 8). Over the past 100 years, however, nest temperatures on both "cool" and "warm" beaches have increased nearly 0.5°C during the nesting season. This study suggests that the worldwide production of green turtles hinges on the future availability of cool beaches (Hays et al. 2003). The physiological effects of increased temperature also depend on how close an organism lives to its upper thermal limit (Figure 1.4). Many reef-building corals, for example, live near their upper limit. A change of just a couple of degrees can result in **bleaching** (i.e., expulsion of symbiotic algae), reduced reproductive success, and even death (Harley et al. 2006).

Figure 3.2

A green turtle returns to the sea after nesting on Ascension Island.

Source: Hays, G. C. © 2004, Elsevier. Modified with permission from Elsevier.

Over the short term, at least, climate change can have positive effects on the physiology of some species. For instance, researchers in southern France examined four populations of common lizards (*Lacerta vivipara*) over 17 years. They found a correlation between rising temperatures and large body size, which in turn increased adult survival and the number of offspring. Lizards, being **ectothermic** (i.e., "cold blooded"), grow more quickly in warm conditions partly because they can spend more time each day looking for food. Large females can produce more eggs because their abdominal cavities are more spacious. Bigger lizards also are more likely to escape predators because they can run faster and for a longer time (Chamaillé-Jammes et al. 2006). Unfortunately, an apparently positive physiological effect for some species can have negative consequences for others. In a review of 75 studies, biologists in Florida found that elevated levels of CO_2 increased the productivity of plant species but decreased survival in insect herbivores (Stiling and Cornelissen 2007). Plants grown under high CO_2 are starchier, have less nitrogen, and produce more toxins than plants grown under normal conditions. Insects feeding on these plants grow more slowly, have lower reproductive success, and seem to be more susceptible to predators. Additionally, because the insects have to consume more plant tissue to meet their nutritional needs, they can cause more overall structural damage to the plants.

For a discussion of the effects of a warming climate on human physiology, please see the open-ended inquiry in Chapter 10.

Phenology

Phenology, or the timing of life cycle events, is the most obvious indicator of a species' response to climate change. Because of the cultural and economic importance of the timing of migration, fruiting, and other events, humans have monitored thousands of species for up to hundreds of years. In one study, 62% of 677 species observed over 16–132 years showed advancement of spring activities (Parmesan and Yohe 2003; see examples in Table 3.1). On average, the growing season in mid- to high latitudes of the Northern Hemisphere has been extended by about two weeks in the last 50 years, due both to an earlier spring and a later fall (Rosenzweig et al. 2007). Your students can explore how changes in plant phenology can affect human health with the activity in Chapter 9.

Because climate change affects each species differently, there is significant potential for **trophic mismatches** (i.e., the phenology of one species no longer matches that of its food source). In a large lake near Seattle, Washington, for example, ecologists have observed a 20-day advance in the timing of phytoplankton blooms since 1977. Although some herbivorous zooplankton have likewise advanced their spring phenology, *Daphnia pulicaria* has not and consequently is suffering population declines. *Daphnia pulicaria* is a critical food source for many fish (Winder and Schindler 2004). The decoupling of species interactions, like those between predators and prey or pollinators and host plants, has more serious implications for ecosystems than a change in one species alone (Parmesan 2006). The activity in Chapter 8 investigates the mismatch between songbirds and their caterpillar prey.

Distribution and Abundance

Throughout Earth's history, species have responded to climate change primarily through shifting their **range** (i.e., geographic distribution) and abundance (Fischlin et al. 2007). For example, spruce (*Picea* spp.) tends to grow in cool, moist habitats. About 21,000 years ago, when eastern North America was dominated by an ice sheet that extended south of the Great Lakes and the Laurentian River Valley, spruce was widespread from the southern edge of the ice margin to east-central Louisiana (Jackson et al. 2000). As the climate warmed and the ice retreated, the range of spruce shifted north. The southern edge of the range also contracted, likely because *Picea critchfieldii*, a species that tolerated warmer conditions, went extinct (Jackson et al. 2000). By the time most of the eastern North American ice sheet had disappeared about 6,000 years ago, spruce occupied only the extreme northeast of the United States to north-central Canada (Jacobson, Webb, and Grimm 1987).

As the climate warms, scientists expect organisms to move upward to higher altitudes and poleward to higher latitudes (Parmesan 2006)—much like spruce during deglaciation. One study reports that of 893 species monitored over the past 17–1,000 years, 80% relocated upward and/or poleward (Parmesan and Yohe 2003).

Changes in distribution are intimately related to the size of populations along the margins of a species' range. A range margin contracts when populations go locally extinct, through processes such as mass mortality events (e.g., hurricanes), reduced survivorship (e.g., because of disease), or reduced reproductive success (e.g., because of insufficient food). For example, rising autumn temperatures may be contracting the southern range of grey jays (*Perisoreus canadensis*; Figure 3.3, p. 56). In late summer, grey jays hoard perishable foods such as berries and insects, relying on cool autumn and winter temperatures to prevent rot. The food sustains the birds through the winter and into their breeding season in early spring. Over the past 26 years, it appears warming autumn weather has led to a loss of food stores and precipitous declines in jay populations (Waite and Strickland 2006).

A range margin expands when organisms establish new populations outside their former range, perhaps because of changes in habitat suitability or the accessibility of food (McCarty 2001). For instance, the winter pine processionary moth (*Thaumetopoea pityocampa*), an economically important forest pest, has expanded over 87 km northward in France and up to 230 m upward in the Italian Alps since the 1970s. The caterpillars feed on the needles of conifers through the winter as long as temperatures exceed 0°C. Warming winters in Europe have expanded the potential habitat for the caterpillars by

Figure 3.3

Mated pair of grey jays in Algonquin Park, Ontario, Canada.

Photograph courtesy of Thomas Waite.

increasing the time available for feeding and by reducing exposure to lethal temperatures (about –16°C) (Battisti et al. 2005). In another example, a new record has been set for the elevation limit of amphibians in the Peruvian Andes. Due to glacial retreat over the last 70 years, three species of frogs and toads now occupy ponds at 5,244 m of elevation and one species occupies ponds at 5,400 m (Seimon et al. 2007). The previous record, reported from Chile in 2003, was 5,000 m.

Climate change acts on biological systems locally, not globally, so it is unlikely to be the sole cause of extinction of most species (McCarty 2001). However, climate change poses an imminent risk for species with small ranges, low population densities within their range, narrow climate tolerances, or lack of alternative habitats. For instance, the Aldabra banded snail (*Rhachistia aldabrae*), which was only known from a few islands in the Indian Ocean, appears to have gone extinct sometime between 1997 and 2000. The decline and eventual extinction of the snail was probably caused by decreasing rainfall. When conditions were dry, snails conserved body water by becoming dormant. Prolonged dormancy left insufficient time to feed, grow, or reproduce, with particularly severe consequences for young snails (Gerlach 2007). Species facing multiple environmental threats are also more likely to go extinct. In an experiment with rotifers, Mora et al. (2007) demonstrated that extinction could be caused by environmental warming alone, but that the rate of extinction was accelerated by harvesting and loss of habitat. Students can use the polar bear activity in Chapter 6 to investigate the effects of climate change on population dynamics and extinction risk.

Ecosystem Structure and Function

The biological responses we've discussed so far have implications for the structure and function of communities and ecosystems. As organisms relocate and change their behavior, **novel ecosystems**—combinations of species that haven't previously occurred within a given biome—are likely to form (Hobbs et al. 2006). Ecologists are uncertain how the reorganization of biological systems will affect ecosystem services important to humans (Table 3.2).

Topic: Ecosystem dynamics
Go to: www.scilinks.org
Code: CCPP12

Table 3.2

- Clean drinking water
- Flooding and storm damage control
- Pollination of crop plants
- Wild-harvested medicines and foods
- Reduction/control of vector-borne diseases
- Nature-based tourism
- Natural resource jobs

Examples of ecosystem services provided by wild (i.e., unmanaged) systems.

Source: Information compiled from Balmford and Bond 2005.

Most of our current understanding of how climate change affects communities and ecosystems comes from short-term, small-scale experiments. However, early observations of natural systems suggest that large-scale biotic reorganizations already have occurred. In the Antarctic, for example, loss of sea ice has altered penguin communities (see the activity in Chapter 5). In the Arctic, expanses of tundra are being replaced by shrubs due to increased microbial activity (i.e., nitrogen availability) in warm winters (Figure 3.4). As the shrubs grow, they create a positive feedback by trapping snow, which insulates the soil and enhances microbial activity. This shrub expansion has the potential to reduce the size of caribou populations because shrubs are less nutritious than grasses and lichens. Perhaps more important, shrub expansion could amplify regional climate warming. Shrubs have a low albedo and higher sensible heat transfer (see Chapter 1) than grasses, lichens, and mosses (Sturm et al. 2005). Finally, thawing permafrost is a carbon source rather than a carbon sink. Because of increased soil temperatures and water availability, more carbon is lost to the atmosphere through decomposition than is gained from the atmosphere by photosynthesis (Oechel et al. 1993).

Figure 3.4

Positive feedback loop responsible for replacement of tundra vegetation by woody shrubs.

Source: Sturm, M. et al. 2005. © 2005, American Institute of Biological Sciences (AIBS). Modified with permission from AIBS.

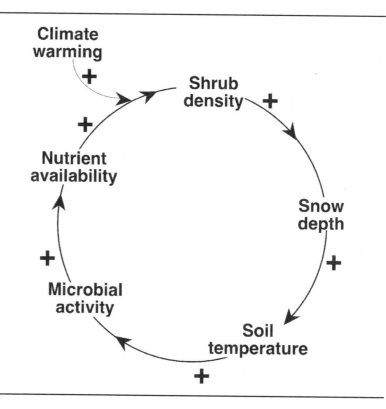

Ecosystems change on multiple time scales, from days to thousands of years. The fluid nature of ecosystems makes it difficult to project long-term change. For example, on the central coast of California where summer drought limits plant growth, precipitation will probably increase under a warming climate. In a multiyear study, researchers enhanced the spring rainy season in outdoor experimental plots. Initially, it appeared that extra rain would simply increase the productivity of nitrogen-fixing herbs and the diversity and abundance of invertebrates. In the second year, however, the growth of annual grasses exploded because of the increase in nitrogen. As the grasses died at the end of each growing season, the accumulated litter suppressed the growth of other plant species and—because grass is less nutritious and structurally complex than a mixed-plant community—dramatically reduced the diversity and abundance of invertebrates (Suttle, Thomsen, and Power 2007).

As we mentioned above, wild species face a number of threats including habitat fragmentation, pollution, and overhunting. **Nonnative** plants and animals (species introduced by people to a new habitat) are considered one of the world's primary conservation threats. These species—also called "exotic," "alien," or "invasive"— can consume, compete with, or even crossbreed with native species, causing local extinctions and altering ecosystem processes such as nutrient cycling (Mack et al. 2000). Both extreme and gradual changes in climate can make ecosystems more susceptible to invasion (Fischlin et al. 2007). For instance, common cordgrass (*Spartina anglica*) is a fertile hybrid of two different cordgrass species. It has been planted in many parts of the world to control coastal erosion. Common cordgrass converts bare mudflats and well-drained marshes to dense, poorly drained monocultures, which tend to reduce the diversity of fish, birds, invertebrates, and other plants (State Noxious Weed Control Board 2007). In Germany, researchers found that common cordgrass has spread north since 1988 because of increasing spring temperatures (the species cannot germinate below 4°C and photosynthesis is slow below 7°C) (Loebl, van Beusekom, and Reise 2006).

Climate change probably will have negative effects on most biological systems. However, some diverse ecosystems, with complex food webs and healthy populations of native species, appear somewhat resilient to climate change. In Yellowstone National Park, for example, the reintroduction of wolves has slowed the negative effects of declining winter snow cover. Many carnivores and scavengers in the park rely on the carcasses of elk that starve when the snow is deep. Because wolves rarely eat an entire kill, leftover **carrion** (rotting flesh) is available for other species, regardless of the weather (Wilmers and Getz 2005). You and your students can explore this topic further with the activity in Chapter 7.

Evolution of New Traits

Biologists have observed a few instances of rapid evolution in response to climate change. In southern California, for instance, a population of field mustard (*Brassica rapa*) evolved earlier flowering times in response to drought in just a few generations (Franks, Sim, and Weis 2007). In northern Canada, red squirrels (*Tamiasciurus hudsonicus*) evolved earlier reproduction times in response to warming springs in just 10 years (Réale et al. 2003).

Unfortunately, evolution is unlikely to be a widespread "fix" for climate change, for the following reasons. First, selection pressures may change too quickly for adaptations to keep up, particularly in long-lived organisms (Barnosky and Kraatz 2007). Species with small bodies, short life cycles, and large populations have the best chance to genetically adapt to climate change (Bradshaw and Holzapfel 2006). Second, populations are more likely to shift their geographical range in search of their preferred climate than to adapt to new conditions (Parmesan 2006). This tendency to retain tolerance to a specific climate can exist in a population because of **stabilizing selection** (i.e., selection against nonoptimal phenotypes) or because of a lack of genetic variation. Also, potentially adaptive genetic changes in a population can sometimes be swamped by genes from other populations, and traits linked to survival of adverse conditions can be linked to other traits that reduce fitness (Wiens and Graham 2005). The fact that so many species have experienced contractions at the low-elevation and/or southerly margin of their ranges suggest that rapid evolution to climate change is simply not very common (Parmesan 2006).

CLIMATE CHANGE IN THE CLASSROOM

Earth's climate is changing at an unprecedented rate, largely because of human activities. Rapid changes in temperature, precipitation, storm intensity, and other climate elements have consequences for multiple levels of biological organization, from DNA to ecosystems and beyond. Researchers from all over the world are working together to refine scientific understanding of how our climate works and how it interacts with living things. Because of its interdisciplinary nature, climate change is a perfect vehicle for addressing multiple aspects of your science curriculum.

The final chapter of this section—"Quick Guide to Climate"—is a brief, student-friendly overview of Chapters 1 through 3. The overview is suitable either for introducing climate change to your students or for reviewing key concepts at the end of a unit. You also can use the overview in part or in entirety as background information for the classroom activities in Part II of the book.

REFERENCES

Balmford, A., and W. Bond. 2005. Trends in the state of nature and their implications for human well-being. *Ecology Letters* 8(11): 1218–1234.

Barnosky, A. D., and B. P. Kraatz. 2007. The role of climatic change in the evolution of mammals. *BioScience* 57(6): 523–532.

Battisti, A., M. Stastny, S. Netherer, C. Robinet, A. Schopf, A. Roques, and S. Larsson. 2005. Expansion of geographic range in the pine processionary moth caused by increased winter temperatures. *Ecological Applications* 15(6): 2084–2096.

Bradshaw, W. E., and C. M. Holzapfel. 2006. Evolutionary response to rapid climate change. *Science* 312(5779): 1477–1478.

Brookhaven National Laboratory (BNL). 2006. The FACE Program. Retrieved October 24, 2007, from *www.bnl.gov/face/faceProgram.asp*

Chamaillé-Jammes, S., M. Massot, P. Aragón, and J. Clobert. 2006. Global warming and positive fitness response in mountain populations of common lizards *Lacerta vivipara*. *Global Change Biology* 12(2): 392–402.

Crozier, L. 2002. Climate change and its effect on species range boundaries: A case study of the sachem skipper butterfly, *Atalopedes campestris*. In *Wildlife responses to climate change: North American case studies*, eds. S. H. Schneider, and T. L. Root, 57–91. Washington, DC: Island Press.

Denman, K. L., G. Brasseur, A. Chidthaisong, P. Ciais, P. M. Cox, R. E. Dickinson, D. Hauglustaine, C. Heinze, E. Holland, D. Jacob, U. Lohmann, S. Ramachandran, P. L. da Silva Dias, S. C. Wofsy, and X. Zhang. 2007. Couplings between changes in the climate system and biogeochemistry. In *Climate change 2007: The physical science basis. Contribution of Working Group I to the Fourth Assessment Report of the Intergovernmental Panel on Climate Change*, eds. S. Solomon, D. Qin, M. Manning, Z. Chen, M. Marquis, K. B. Averyt, M. Tignor, and H. L. Miller, 499–588. Cambridge: Cambridge University Press.

Dunne, J. A., S. R. Saleska, M. L. Fischer, and J. Harte. 2004. Integrating experimental and gradient methods in ecological climate change research. *Ecology* 85(4): 904–916.

Fischlin, A., G. F. Midgley, J. T. Price, R. Leemans, B. Gopal, C. Turley, M. D. A. Rounsevell, O. P. Dube, J. Tarazona, and A. A. Velichko. 2007. Ecosystems, their properties, goods and services. In *Climate change 2007: Impacts, adaptation and vulnerability. Contribution of Working Group II to the Fourth Assessment Report of the Intergovernmental Panel on Climate Change*, eds. M. L. Parry, O. F. Canziani, J. P. Palutikof, P. J. van der Linden, and C. E. Hanson, 211–272. Cambridge: Cambridge University Press.

Franks, S. J., S. Sim, and A. E. Weis. 2007. Rapid evolution of flowering time by an annual plant in response to a climate fluctuation. *Proceedings of the National Academy of Sciences of the United States of America* 104(4): 1278–1282.

Gerlach, J. 2007. Short-term climate change and the extinction of the snail *Rhachistia aldabrae* (Gastropoda: Pulmonata). *Biology Letters* 3(5): 581–584.

Harley, C. D. G., A. R. Hughes, K. M. Hultgren, B. G. Miner, C. J. B. Sorte, C. S. Thornber, L. F. Rodriguez, L. Tomanek, and S. L. Williams. 2006. The impacts of climate change in coastal marine systems. *Ecology Letters* 9(2): 228–241.

Hays, G. C. 2004. Good news for sea turtles. *Trends in Ecology and Evolution* 19(7): 349–351.

Hays, G. C., A. C. Broderick, F. Glen, and B. J. Godley. 2003. Climate change and sea turtles: A 150-year reconstruction of incubation temperatures at a major marine turtle rookery. *Global Change Biology* 9(4): 642–646.

Helmuth, B., J. G. Kingsolver, and E. Carrington. 2005. Biophysics, physiological ecology, and climate change: Does mechanism matter? *Annual Review of Physiology* 67: 177–201.

Hobbs, R. J., S. Arico, J. Aronson, J. S. Baron, P. Bridgewater, V. A. Cramer, P. R. Epstein, J. J. Ewel, C. A. Klink, A. E. Lugo, D. Norton, D. Ojima, D. M. Richardson, E. W. Sanderson, F. Valladares, M. Vilà, R. Zamora, and M. Zobel. 2006. Novel ecosystems: Theoretical and management aspects of the new ecological world order. *Global Ecology and Biogeography* 15(1): 1–7.

Huntington, H., T. Callaghan, S. Fox, and I. Krupnik. 2004. Matching traditional and scientific observations to detect environmental change: A discussion on Arctic terrestrial ecosystems. *Ambio* Special Report 13: 18–23.

Jackson, S. T., R. S. Webb, K. H. Anderson, J. T. Overpeck, T. Webb, J. W. Williams, B. C. S. Hansen. 2000. Vegetation and environment in eastern North America during the Last Glacial Maximum. *Quaternary Science Reviews* 19(6): 489–508.

Jackson, S. T. and C. Weng. 1999. Late Quaternary extinction of a tree species in eastern North America. *Proceedings of the National Academy of Sciences of the United States of America* 96 (24): 13847–13852.

Jackson, S. T., and J. W. Williams. 2004. Modern analogs in Quaternary paleoecology: Here today, gone yesterday, gone tomorrow? *Annual Review of Earth and Planetary Sciences* 32: 495–537.

Jacobson, G. L., T. Webb, and E. C. Grimm. 1987. Patterns and rates of vegetation change during the deglaciation of eastern North America. In *North America and adjacent oceans during the last glaciation*, eds. W. F. Ruddiman, and H. E. Wright, 277–288. Boulder, CO: Geological Society of America.

Kerr, J. T., H. M. Kharouba, and D. J. Currie. 2007. The macroecological contribution to global change solutions. *Science* 316(5831): 1581–1584.

Laidler, G. J. 2006. Inuit and scientific perspectives on the relationship between sea ice and climate change: The ideal complement? *Climatic Change* 78 (2–4): 407–444.

Loebl, M., J. E. E. van Beusekom, and K. Reise. 2006. Is spread of the neophyte *Spartina anglica* recently enhanced by increasing temperatures? *Aquatic Ecology* 40(3): 315–324.

Logan, J. A., J. Régnière, D. R. Gray, and A. S. Munson. 2007. Risk assessment in the face of a changing environment: Gypsy moth and climate change in Utah. *Ecological Applications* 17(1): 101–117.

Mack, R. N., D. Simberloff, W. M. Lonsdale, H. Evans, M. Clout, and F. A. Bazzaz. 2000. Biotic invasions: Causes, epidemiology, global consequences, and control. *Ecological Applications* 10(3): 689–710.

McCarty, J. P. 2001. Ecological consequences of recent climate change. *Conservation Biology* 15(2): 320–331.

Mora, C., R. Metzger, A. Rollo, and R. A. Myers. 2007. Experimental simulations about the effects of overexploitation and habitat fragmentation on populations facing environmental warming. *Proceedings of the Royal Society B* 274(1613): 1023–1028.

Morrison, L. W., M. D. Korzukhin, and S. D. Porter. 2005. Predicted range expansion of the invasive fire ant, *Solenopsis invicta*, in the eastern United States based on the VEMAP global warming scenario. *Diversity and Distributions* 11(3): 199–204.

Oechel, W. C., S. J. Hastings, G. Vourlitis, M. Jenkins, G. Riechers, and N. Grulke. 1993. Recent change of Arctic tundra ecosystems from a net carbon dioxide sink to a source. *Nature* 361(6412): 520–523.

Parmesan, C. 2006. Ecological and evolutionary responses to recent climate change. *Annual Review of Ecology, Evolution, and Systematics* 37: 637–669.

Parmesan, C., and G. Yohe. 2003. A globally coherent fingerprint of climate change impacts across natural systems. *Nature* 421(6918): 37–42.

Réale, D., D. Berteaux, A. G. McAdam, and S. Boutin. 2003. Lifetime selection on heritable life-history traits in a natural population of red squirrels. *Evolution* 57(10): 2416–2423.

Rosenzweig, C., G. Casassa, D. J. Karoly, A. Imeson, C. Liu, A. Menzel, S. Rawlins, T. L. Root, B. Seguin, and P. Tryjanowski. 2007. Assessment of observed changes and responses in natural and managed systems. In *Climate change 2007: Impacts, adaptation and vulnerability. Contribution of Working Group II to the Fourth Assessment Report of the Intergovernmental Panel on Climate Change*, eds. M. L. Parry, O. F. Canziani, J. P. Palutikof, P. J. van der Linden, and C. E. Hanson, 79–131. Cambridge: Cambridge University Press.

Seimon, T. A., A. Seimon, P. Daszak, S. R. P. Halloy, L. M. Schloegel, C. A. Aguilar, P. Sowell, A. D. Hyatt, B. Konecky, and J. E. Simmons. 2007. Upward range extension of Andean anurans and chytridiomycosis to extreme elevations in response to tropical deglaciation. *Global Change Biology* 13(1): 288–299.

Singer, F. 2007. Dualism, science, and statistics. *BioScience* 57(9): 778–782.

State Noxious Weed Control Board. 2007. Common cordgrass (*Spartina anglica* C. E. Hubbard). Retrieved October 1, 2007, from *www.nwcb.wa.gov/weed_info/written_findings/Spartina_anglica.html*

Stiling, P., and T. Cornelissen. 2007. How does elevated carbon dioxide (CO_2) affect plant-herbivore interactions? A field experiment and meta-analysis of CO_2-mediated changes on plant chemistry and herbivore performance. *Global Change Biology* 13(9): 1823–1842.

Sturm, M., J. Schimel, G. Michaelson, J. M. Welker, S. F. Oberbauer, G. E. Liston, J. Fahnestock, and V. E. Romanovsky. 2005. Winter biological processes could help convert arctic tundra to shrubland. *BioScience* 55(1): 17–26.

Suttle, K. B., M. A. Thomsen, and M. E. Power. 2007. Species interactions reverse grassland responses to changing climate. *Science* 315(5812): 640–642.

Thuiller, W. 2007. Climate change and the ecologist. *Nature* 448(7153): 550–552.

Waite, T. A., and D. Strickland. 2006. Climate change and the demographic demise of a hoarding bird living on the edge. *Proceedings of the Royal Society B* 273(1603): 2809–2813.

Wiens, J. J., and C. H. Graham. 2005. Niche conservatism: Integrating evolution, ecology, and conservation biology. *Annual Review of Ecology, Evolution, and Systematics* 36: 519–539.

Wilmers, C. C., and W. M. Getz. 2005. Gray wolves as climate change buffers in Yellowstone. *PLoS Biology* 3(4): 0571–0576.

Winder, M., and D. E. Schindler. 2004. Climate change uncouples trophic interactions in an aquatic ecosystem. *Ecology* 85(8): 2100–2106.

Chapter 4
Quick Guide to Climate

HOW EARTH'S CLIMATE WORKS

- **Climate** is the state of the atmosphere over years or decades.

- Earth's climate is controlled by a complex, interactive system composed of land, water, snow and ice, organisms, and the atmosphere. The Sun powers this system.

- About 70% of **shortwave radiation** (visible light) that reaches the atmosphere is absorbed by the climate system and converted to **longwave radiation** (far infrared radiation), eventually being released back into space.

- The natural **greenhouse effect** occurs because certain gases in the atmosphere (e.g., carbon dioxide, water vapor, and methane) allow shortwave radiation from the Sun to enter the atmosphere but block longwave radiation from leaving the atmosphere. Eventually, all longwave radiation escapes into space.

- The temporary absorption of energy by the atmosphere makes life on Earth possible. Without the greenhouse effect, the planet would be about 33°C cooler than it is now.

- Since air and water are fluids that tend to move when heated, the Sun also drives their circulation. This circulation happens in predictable currents and affects temperature and precipitation worldwide.

- Earth's climate system strongly influences living things. The temperature and precipitation of a region determine which organisms can live there. Over long periods, some species can adapt to changes in climate.

HOW EARTH'S CLIMATE IS CHANGING

- To find patterns in climate change, scientists use

  - **Historical sources**—Human records like cave paintings, farming diaries, and ship logs can help reconstruct *past* climate.
  - **Proxies**—Things like pollen, fossils of corals, tree rings, sediment in lakes, and ice cores also can help reconstruct *past* climate.
  - **Instrumental measurements**—Global measurements of temperature, precipitation, and gas concentrations can describe climate from the *recent past* and *present day*.
  - **Modeling**—Mathematical models can be used to forecast *future* climate. No model is perfect, but the knowledge and technology on which models are based is getting better every day.

- Earth's climate is incredibly complex, so scientists are **uncertain** about many aspects of climate change.

- Despite this uncertainty, *scientists overwhelmingly agree that Earth's climate is changing at an alarming rate and magnitude, due in part to human activity.*

  - Since 1906, Earth's average surface temperature has increased by 0.7°C (Figure 4.1). This rate of warming is

Figure 4.1

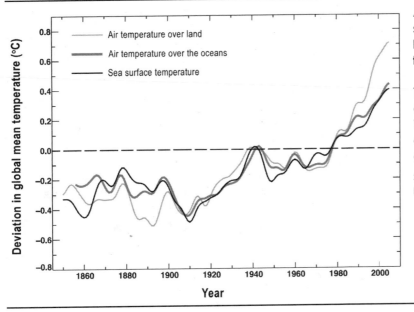

Annual deviation of global air and sea surface temperatures from the long-term mean (average of all years from 1961–1990).

Temperatures at the dotted line are equal to the long-term mean. Warmer-than-average years appear above the line, whereas cooler-than-average years appear below the line.

Source: Trenberth et al. 2007. Modified with permission.

Figure 4.2

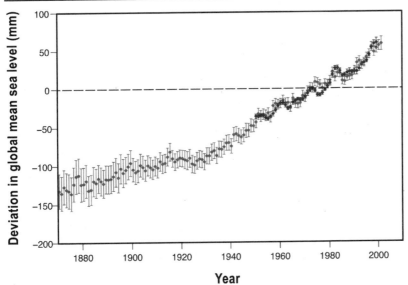

Annual deviation of global sea level (mm) from the long-term mean (average of all years from 1961–1990).

Years with higher-than-average sea level appear above the dotted reference line, whereas years with lower-than-average sea level appear below the line. Before 1950, there were fewer than 100 tide gauges around the world. Therefore, the light gray curve shows sea levels reconstructed from tide gauges and models that include modern-day satellite data. The dark gray curve shows coastal tide gauge measurements since 1950. The close match between the reconstructed and observed sea levels gives scientists a high level of confidence that the models are representative of years prior to 1950.

Source: Bindoff et al. 2007. Modified with permission.

up to nine times faster than at the end of the last ice age.
- Droughts have become more frequent and severe worldwide.
- The number of severe hurricanes has increased worldwide.
- Snow, sea ice, mountain glaciers, and ice caps are melting worldwide.
- Sea levels are increasing worldwide (Figure 4.2).
- The concentration of CO_2 is increasing in the oceans, making them more acidic.

- While some of these changes may be unrelated to human activity, the warming of the last 50 years has been caused mostly by humans. About 75% of the warming has been caused by increases in the **greenhouse gases** carbon dioxide (Figure 4.3) and methane.

Figure 4.3

Trend in atmospheric concentration of carbon dioxide from the last two thousand years.

Source: Forster et al. 2007. Modified with permission.

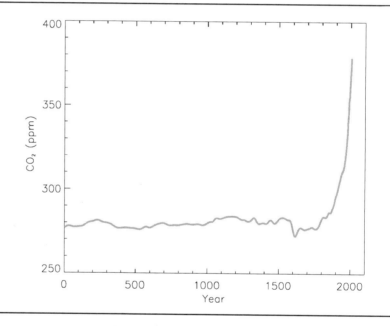

- **Positive feedbacks** are cycles that intensify the warming caused by extra greenhouse gases. For example, the more ice melts due to warm air, the less radiation is reflected by ice back into space. This means more energy is absorbed by the atmosphere, meaning more warming, more melting, and so on.

HOW CLIMATE CHANGE IS AFFECTING LIVING THINGS

- Scientists use observational studies, experiments, and modeling to measure the effects of climate change on living things and to predict future effects.

    - Climate change can affect the physiology and morphology (life processes and body form) of some organisms. The effects can be positive or negative. Some species, for example, may grow more quickly and get bigger, whereas other species may sicken and die because of heat or drought.
    - Climate change can change the timing, or **phenology**, of some life processes, including flowering, ripening, migration, reproduction, and hibernation. In some tree species, for example, buds can open earlier in the spring and leaves can drop later in the fall. These phenological changes can be disastrous for some species.
    - Climate change can affect where populations live and how many individuals there are in each population. A population's **range** (the area in which it lives) can expand, contract, or shift to a new area. The size of a population may increase or decrease.
    - Climate change can affect whole ecosystems. It can cause native species to be lost from one area and be replaced by **non-native species** that don't belong in that area. Climate change also can alter the productivity of plants and how organisms in a food web interact with one another.

- Some populations may be able to evolve in response to climate change, especially when the populations are large and the individuals in the population have a short life span. However, *widespread rapid evolution is not common* and is unlikely to provide a biological solution to climate change.

FREQUENTLY ASKED QUESTIONS

Q: *How do scientists know what climate was like 650,000 years ago?*
A: Scientists use proxies—which are things like fossils, air bubbles

in glaciers, and mud from the bottom of the ocean—to reconstruct past climates. Proxies can provide information from climates as recent as a few hundred years ago to over a million years ago. Because some proxies are more informative or accurate than others, scientists usually use multiple proxies to increase confidence in their results.

Q: *Couldn't the warming we see now just be a natural cycle in global temperature?*

A: It is highly unlikely that the warming we see now is a natural cycle. The Earth is warmer now than it has been in at least the last 1,000 years. The warming trend is up to nine times faster than at the end of the last ice age. Glaciers are retreating faster than in the last 12,000 years. The sea level is rising faster than in the last 2–3 thousand years. Finally, there's more carbon dioxide and methane in the atmosphere than at any other time in the last 650,000 years (Figure 4.4). Scientists think that without human input of greenhouse gases in the atmosphere, Earth probably would have cooled slightly over the last 50 years.

Figure 4.4

Variations in atmospheric CO_2, atmospheric CH_4, and deuterium (δD), as derived from air bubbles in Antarctic ice cores.

There is a linear relationship between δD, which is an isotope of hydrogen, and local temperature. The gray vertical bars indicate interglacial warm periods of the last 450,000 years.

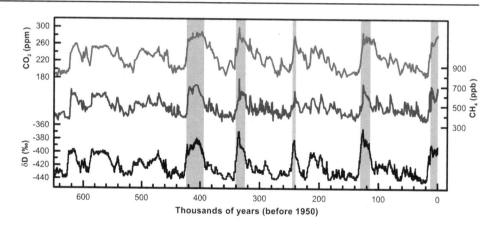

Source: Jansen et al. 2007. Modified with permission.

Q: *Why is less than 1°C of warming such a big deal?*

A: Remember that the entire planet is warming by an average of 0.7°C. In some parts of the world, temperatures have increased by as much as 3–4°C (the difference between sweater weather and t-shirt weather). Also, other changes in climate, like increases in droughts and hurricanes, accompany increasing temperatures. Organisms can survive, grow, and reproduce only within a limited range of physical conditions. Even if an organism is not sensitive to a small change in temperature or rainfall, it probably depends on another organism that is. Ecosystems provide humans with clean water, food, building materials, and protection from storms, diseases, and pests. Just a few degrees of warming can disrupt these services.

Q: *Can't organisms just adapt to a warmer climate?*

A: Most organisms probably can't adapt to the current changes in climate. Natural selection typically occurs slowly (over thousands to hundreds of thousands of years) but the climate is changing rapidly (over decades). Also, when species are faced with a changing climate, they usually try to move to a new area that still has their preferred climate. If most individuals of a species can successfully move, there's no pressure to adapt to the new climate.

REFERENCES

Bindoff, N. L., J. Willebrand, V. Artale, A, Cazenave, J. Gregory, S. Gulev, K. Hanawa, C. Le Quéré, S. Levitus, Y. Nojiri, C. K. Shum, L. D. Talley, and A. Unnikrishnan. 2007. Observations: Oceanic climate change and sea level. In *Climate change 2007: The physical science basis. Contribution of Working Group I to the Fourth Assessment Report of the Intergovernmental Panel on Climate Change*, eds. S. Solomon, D. Qin, M. Manning, Z. Chen, M. Marquis, K. B. Averyt, M. Tignor, and H. L. Miller, 385–432. Cambridge: Cambridge University Press.

Forster, P., V. Ramaswamy, P. Artaxo, T. Berntsen, R. Betts, D. W. Fahey, J. Haywood, J. Lean, D. C. Lowe, G. Myhre, J. Nganga, R. Prinn, G. Raga, M. Schulz, and R. Van Dorland. 2007. Changes in atmospheric constituents and in radiative forcing. In *Climate change 2007: The physical science basis. Contribution of Working Group I to the Fourth Assessment Report of the Intergovernmental Panel on Climate Change*, eds. S. Solomon, D. Qin, M. Manning, Z. Chen, M. Marquis, K. B. Averyt, M. Tignor, and H. L. Miller, 129–234. Cambridge: Cambridge University Press.

Jansen, E., J. Overpeck, K. R. Briffa, J.-C. Duplessy, F. Joos, V. Masson-Delmotte, D. Olago, B. Otto-Bliesner, W. R. Peltier, S. Rahmstorf, R. Ramesh, D. Raynaud, D. Rind, O. Solomina, R. Villalba, and D. Zhang. 2007. Palaeoclimate. In *Climate change 2007: The physical science basis. Contribution of Working Group I to the Fourth Assessment Report of the Intergovernmental Panel on Climate Change,* eds. S. Solomon, D. Qin, M. Manning, Z. Chen, M. Marquis, K. B. Averyt, M. Tignor, and H. L. Miller, 433–497. Cambridge: Cambridge University Press.

Trenberth, K. E., P. D. Jones, P. Ambenje, R. Bojariu, D. Easterling, A. Klein Tank, D. Parker, F. Rahimzadeh, J. A. Renwick, M. Rusticucci, B. Soden, and P. Zhai. 2007. Observations: Surface and atmospheric climate change. In *Climate change 2007: The physical science basis. Contribution of Working Group I to the Fourth Assessment Report of the Intergovernmental Panel on Climate Change,* eds. S. Solomon, D. Qin, M. Manning, Z. Chen, M. Marquis, K. B. Averyt, M. Tignor, and H. L. Miller, 235–336. Cambridge: Cambridge University Press.

Part II: Climate Change Case Studies

CONNECTIONS TO STANDARDS

In this section of the book, we present six classroom investigations that use the case study method. The use of real-life stories makes science more understandable and engaging to students. Furthermore, open-ended cases illustrate the nature of science better than cookbook "experiments" with a predetermined right answer (Herreid 2007). In this collection of investigations, students analyze data from recent scientific research, solve realistic problems, and communicate their interpretations and decisions to their peers in a variety of ways.

As we discussed in Chapter 3, scientists have detected the effects of anthropogenic climate change in nearly every taxa and in all biomes. To engage student interest, however, we focused on charismatic species such as penguins and polar bears. Because relatively little research has been done on the effects of climate change in South America, Africa, and Asia (Parmesan 2006), most of the case studies are from North America and Europe.

Each case study corresponds to multiple National Science Education Standards (NRC 1996). You can use the table on the next page to help you choose the case most relevant to your instructional needs.

REFERENCES

Herreid, C. F., ed. 2007. *Start with a story: The case study method of teaching college science.* Arlington, VA: NSTA Press.

National Research Council (NRC). 1996. *National science education standards.* Washington, DC: National Academy Press.

Parmesan, C. 2006. Ecological and evolutionary responses to recent climate change. *Annual Review of Ecology, Evolution, and Systematics* 37: 637–669.

Connections to the National Science Education Standards

◆ Primary focus of lesson; ◈ Secondary focus of lesson

National Science Education Standards	CH. 5: Now You "Sea" Ice, Now You Don't	CH. 6: Population Peril	CH. 7: Carrion: It's What's for Dinner	CH. 8: Right Place, Wrong Time	CH. 9: Ah-Choo!	CH. 10: Cruel, Cruel Summer
Content Standard A: Science as Inquiry						
Design and conduct scientific investigations						◆
Use technology and mathematics to improve investigations and communications					◆	◆
Formulate and revise scientific explanations and models using logic and evidence	◆	◆	◆	◆		◆
Recognize and analyze alternative explanations and models	◆	◆		◆		◆
Communicate and defend a scientific argument	◆	◆	◈	◆		◆
Understand that mathematics is essential in scientific inquiry			◆			◆
Content Standard C: Life Sciences						
Biological evolution	◈	◈		◈		
The interdependence of organisms	◆	◆	◆	◆		
Matter, energy, and organization in living systems	◆	◆	◆	◈		
Behavior of organisms		◈		◆		◈
Content Standard E: Abilities of Technological Design						
Propose designs and choose between alternative solutions					◆	
Implement a proposed solution					◆	
Communicate the problem, process, and solution					◆	◈
Understand that many scientific investigations require contributions of individuals from different disciplines	◆					
Content Standard F: Science in Personal and Social Perspectives						
Personal and community health					◆	◆
Natural and human-induced hazards					◈	◆
Science should inform active debate about how to resolve certain social challenges, but cannot resolve challenges alone		◆				◈
Humans have a major effect on other species	◈	◆	◆			
Content Standard G: History and Nature of Science						
Scientists have ethical traditions					◈	
Scientific explanations must meet certain criteria	◆	◆	◆	◆	◆	◆

Chapter 5
Now You "Sea" Ice, Now You Don't

Penguin communities shift on the Antarctic Peninsula

Teacher Pages

AT A GLANCE

Increasing air temperatures in the last 50 years have dramatically altered the Antarctic Peninsula ecosystem. In this interdisciplinary inquiry, learners use a cooperative approach to investigate changes in the living and nonliving resources of the Peninsula. The activity stresses the importance of evidence in the formulation of scientific explanations. (Class time: 1–3.5 hours)

This chapter has been modified from the following article: Constible, J., L. Sandro, and R. E. Lee. 2007. A cooperative classroom investigation of climate change. *The Science Teacher* 74(6): 56–63.

"For many, [the Antarctic Peninsula] is the most beautiful part of the Antarctic, unlocked each year by the retreating ice…. It is on this rocky backbone stretching north that most of the continent's wildlife survives…. Almost every patch of accessible bare rock is covered in a penguin colony. Even tiny crags that pierce the mountainsides are used by nesting birds."
—Alastair Fothergill, *A Natural History of the Antarctic: Life in the Freezer*, 1995

INTRODUCTION

At the global level, strong evidence suggests that observed changes in Earth's climate are largely due to human activities (IPCC 2007). At the regional level, the evidence for human-dominated change is sometimes less clear. Scientists have a particularly difficult time explaining warming trends in Antarctica—a region with a relatively short history of scientific observation and a highly variable climate (Clarke et al. 2007). Regardless of the mechanism of warming, however, climate change is having a dramatic impact on Antarctic ecosystems.

By the end of this lesson, students should be able to do the following:
- Graphically represent data.
- Use multiple lines of evidence to generate scientific explanations of ecosystem-level changes on the Antarctic Peninsula.
- Describe ways in which climate change on the Antarctic Peninsula has led to interconnected, ecosystem-level effects.
- Participate in an interdisciplinary scientific investigation, demonstrating the collaborative nature of science.

WARMING CLIMATE, WANING SEA ICE

Air temperature data indicate that the western Antarctic Peninsula (Figure 5.4, p. 82) has warmed by about 3°C in the last century (Clarke et al. 2007). Although this relatively short-term record is only from a few research stations, other indirect lines of evidence confirm the trend. The most striking of these proxies is a shift in penguin communities. Adélie penguins, which are dependent on sea ice for their survival, are rapidly declining on the Antarctic Peninsula despite a 600-year colonization history. In contrast, chinstrap penguins, which prefer open water, are dramatically increasing (Figure 5.1). These shifts in penguin populations appear to be the result of a decrease in the amount, timing, and duration of sea ice (Figure 5.2).

Figure 5.1

Adélie penguin   Chinstrap penguin

Adélie and chinstrap penguins.
Adélie penguins (*Pygoscelis adeliae*) breed on the coast of Antarctica and surrounding islands. They are named after the wife of French explorer Jules Sébastien Dumont d'Urville. Adult Adélies stand 70–75 cm tall and weigh up to 5 kg.

Chinstrap penguins (*Pygoscelis antarctica*) are primarily found on the Antarctic Peninsula and in the Scotia Arc, a chain of islands between the tip of South America and the Peninsula. Their name comes from the black band running across their chins. Adult chinstraps stand 71–76 cm tall and weigh up to 5 kg.

Photographs courtesy of Michael Elnitsky.

Figure 5.2

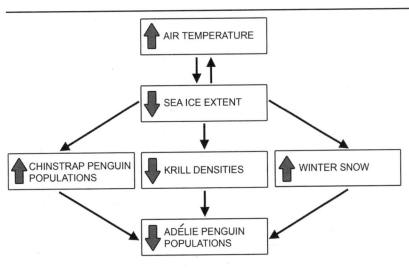

Effects of climate change on sea ice, krill, and penguin communities of the Antarctic Peninsula.

Why is sea ice so important to Adélie penguins? First, sea ice is a feeding platform for Adélies. Krill, the primary prey of Adélies on the Peninsula, feed on microorganisms growing on the underside of the ice (Atkinson et al. 2004). For Adélie penguins, which are relatively slow swimmers, it is easier to find food under the ice than in large stretches of open water (Ainley 2002). Second, sea ice helps control the local climate. Ice keeps the Peninsula cool by reflecting solar radiation back to space. As air temperatures increase and sea ice melts, open water releases heat and amplifies the upward trend in local air temperature (Figure 5.3) (Wadhams 2000). Finally, ice acts as a giant cap on the ocean, limiting evaporation. As sea ice declines, **condensation nuclei** (aerosols that form the core of cloud droplets) and moisture are released into the atmosphere, leading to more snow. This extra snow often does not melt until Adélies have already started nesting; the resulting melt water can kill their eggs (Fraser and Patterson 1997).

Figure 5.3

Melting sea ice amplifies the effects of climate change.

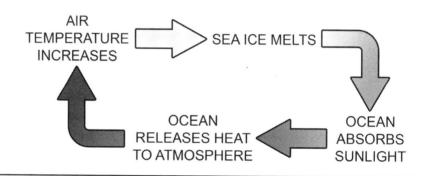

Activity Overview

This directed inquiry uses the jigsaw technique, which requires every student within a group to be an active and equal participant for the rest of the group to succeed (Colburn 2003). To begin, students are organized into "Home Groups" composed of five different specialists. Specialists from each Home Group then reorganize into "Specialist Groups" that contain only one type of scientist (e.g., Group 1 could include all of the Ornithologists and Group 2 all of the Oceanographers). Each Specialist Group receives a piece of the flowchart in Figure 5.2, in the form of a data table. With only a few facts to guide them, the Specialist Groups create graphs from the data tables, brainstorm explanations for patterns in their data, and report results back to their Home Groups. Finally, Home

Groups use the expertise of each specialist to reconstruct the entire flowchart (Figure 5.2).

TEACHING NOTES

Prior Knowledge

Before starting this activity, students should have at least a rudimentary knowledge of Antarctica. You can find a collection of links to our favorite Antarctic websites at *www.units.muohio.edu/cryolab/ education/AntarcticLinks.htm*. You can also engage student interest in this inquiry by showing video clips of penguins, which are naturally appealing to students of all ages. We have short movies of Adélies feeding their young and battling predators on our website at *www.units.muohio.edu/cryolab/education/antarcticbestiary.htm*, and National Geographic has a video called "Rocky Parenting" at *http:// news.nationalgeographic.com/news/2006/11/061117-adelie-video.html*.

Materials

- Specialist Fact Sheet (Student Page 5.1; one for each student, or one overhead for the entire class)
- Temperature data (Figure 5.4; one overhead for the entire class)
- Data sets for each Specialist Group (Student Pages 5.2–5.6: Adélie Penguins, Sea Ice, Winter Snow, Chinstrap Penguins, and Krill)
- Specialist Group Report Sheets (Student Page 5.7; one for each student)
- Sheets of graph paper (one for each student) or computers connected to a printer (one for each Specialist Group)
- Sets of six flowchart cards (one complete set for each Home Group; before the inquiry, you can make flowchart cards by photocopying Figure 5.2 and cutting out each box [i.e., "Air Temperature," "Sea Ice Extent," etc.])
- Paper, markers, and tape for constructing flow charts

Procedure: Graphing and Interpretation

1. Split the class into Home Groups of at least five students each. (Optional: Assign the name of a different real-life research

agency to each group. See *www.units.muohio.edu/cryolab/ education/AntarcticLinks.htm#NtnlProg* for examples.)

2. Instruct students to read the Specialist Fact Sheets (Student Page 5.1). Within a Home Group, each student should assume the identity of a different scientist from the list.

3. Introduce yourself: "Welcome! I'm a climatologist with the Palmer Station, Antarctica Long-Term Ecological Research project. In other words, I study long-term patterns in climate. My colleagues and I have tracked changes in air temperatures on the peninsula since 1947. We have observed that although temperature cycles up and down, it has increased overall [show Figure 5.4]. We think this is occurring because of an increase of greenhouse gases, but we are unsure of the impacts on the Antarctic ecosystem. Your team's job is to describe the interconnected effects of warming on Antarctica's living and nonliving systems."

Figure 5.4

Climatologists: Air temperature data set.

Source: Data compiled from the Palmer Station, Antarctica Long-Term Ecological Research (LTER) data archive. Data from the Palmer LTER archive were supported by the Office of Polar Programs, NSF Grants OPP-9011927, OPP-9632763, and OPP-021782.

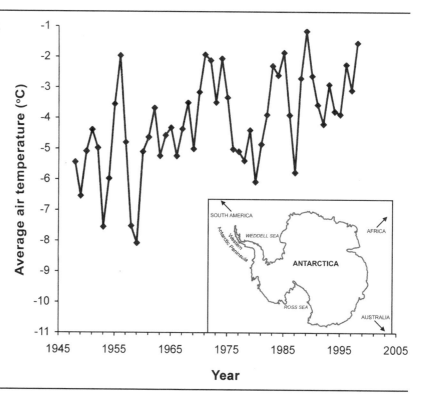

4. Direct the specialists to meet with their respective Specialist Groups. Specialist Groups should not interact with one another.

5. Distribute the data sets and Specialist Group Report Sheets (Student Pages 5.2–5.7) to each Specialist Group. The specialists should graph their data set and interpret the graph.

Procedure: Flowchart and Class Discussion

1. Reconvene the Home Groups.
2. Hand out a complete set of flowchart cards to each group. Each specialist should make a brief presentation to his or her Home Group approximating the format on the Specialist Group Report Sheet (Student Page 5.7). Home Groups should then construct their own flowcharts using all of the flowchart cards. Remind the students throughout this process that they should use the weight of evidence to construct the flowcharts. In other words, each idea should be accepted or rejected based on the amount of support it has.
3. Consider these discussion questions during the flowchart process (do this as a class, by Home Group, or as homework for each student):
   • How has the ecosystem of the Antarctic Peninsula changed in the last 50 years? What are the most likely explanations for these changes?

Figure 5.5

Performance rubric.

Student Name:

Criteria	Points	Self	Teacher	Comments
Active participation in the group process.	5			
Appropriate graph is used to display data. All required elements (labels, titles, etc.) are present. Data are graphed accurately.	5			
Data and interpretations from Specialist Groups are clearly communicated to Home Groups by individual specialists.	10			
Alternative explanations are weighed based on available evidence and prior scientific knowledge.	10			
Conclusions are clearly and logically communicated.	10			
Report Sheet is complete.	5			
TOTAL	**45**			

- Is there sufficient evidence to support these explanations? Why or why not? What further questions are left unanswered?
- Did your Specialist Group come up with any explanations that you think are not very likely (or not even possible!), based on the complete story presented by your Home Group?

Assessment

To assess student learning, you can use a simple performance rubric that focuses on group work and the nature of science (Figure 5.5, p. 83). Depending on the unit of study in which this inquiry is used, a variety of specific content standards also may be assessed. In an ecology unit, for example, you could determine student knowledge of interactions between populations and their environments; in an Earth science unit, you could check student understandings about weather and climate.

Modifications

Some students have initial difficulties with the construction and interpretation of flowcharts. Once students have connected their flowchart cards with arrows, it may be useful to have them label each arrow with a verb. For instance:

For lower-level students, you can construct a worksheet with a "skeleton" of the worksheet (e.g., the general shape of the flowchart and some of the text within the boxes).

You can shorten this lesson by starting immediately with Specialist Groups, rather than with Home Groups. Another option is to provide premade graphs of the data rather than having Specialist Groups create their own.

To make this lesson more open-ended, students may do additional research on the connections between sea ice, krill, and penguins. Note, however, that the majority of resources on this topic are research articles in scientific journals. If you have access to a university library, you might wish to make a classroom file of related journal articles. A more engaging extension would be for students to generate ideas for new research studies that would address questions left unanswered by the current inquiry. This type of activity could range from asking students to formulate new hypotheses to asking students to write short proposals that include specific research questions and plans to answer those hypotheses.

CONCLUSION

Many students have trouble comprehending how just a few degrees of atmospheric warming (in this case, 3°C) could make a difference in their lives. The decline of a charismatic species such as the Adélie penguin is an example of how a seemingly minor change in climate can pose a major threat to plants and animals. Beyond the effects of climate change, however, the activity illustrates the multidisciplinary, international, and, above all, cooperative nature of science. We want social teenagers to realize that they do not have to sit alone in a lab to do science.

REFERENCES

Ainley, D. G. 2002. *The Adélie penguin: Bellwether of climate change.* New York: Columbia University Press.

Atkinson, A., V. Siegel, E. Pakhomov, and P. Rothery. 2004. Long-term decline in krill stock and increase in salps within the Southern Ocean. *Nature* 432(7013): 100–103.

Clarke, A., E. J. Murphy, M. P. Meredith, J. C. King, L. S. Peck, D. K. A. Barnes, and R. C. Smith. 2007. Climate change and the marine ecosystem of the western Antarctic Peninsula. *Philosophical Transactions of the Royal Society B* 362(1477): 149–166.

Colburn, A. 2003. *The lingo of learning: 88 education terms every science teacher should know.* Arlington, VA: NSTA Press.

Fraser, W. R., and D. L. Patterson. 1997. Human disturbance and long-term changes in Adélie penguin populations: A natural experiment at Palmer Station, Antarctic Peninsula. In *Antarctic communities: Species, structure and survival,* eds. B. Battaglia, J. Valencia, and D. W. H. Walton, 445–452. Cambridge: Cambridge University Press.

Intergovernmental Panel on Climate Change (IPCC). 2007. Summary for policymakers. In *Climate change 2007: The physical science basis. Contribution of Working Group I to the Fourth Assessment Report of the Intergovernmental Panel on Climate Change,* eds. S. Solomon, D. Qin, M. Manning, Z. Chen, M. Marquis, K. B. Averyt, M. Tignor, and H. L. Miller, 1–18. Cambridge: Cambridge University Press.

Smith, R. C., W. R. Fraser, and S. E. Stammerjohn. 2003. Climate variability and ecological response of the marine ecosystem in the Western Antarctic Peninsula (WAP) region. In *Climate variability and ecosystem response at Long-Term Ecological Research sites,* eds. D. Greenland, D. G. Goodin, and R. C. Smith, 158–173. New York: Oxford Press.

Wadhams, P. 2000. *Ice in the ocean.* The Netherlands: Gordon and Breach Science Publishers.

OTHER RECOMMENDED RESOURCES

These additional resources were used to create the Student Pages, but are not cited in the text:

Carlini, A. R., N. R. Coria, M. M. Santos, and S. M. Buján. 2005. The effect of chinstrap penguins on the breeding performance of Adélie penguins. *Folia Zoologica* 54(1–2): 147–158.

Forcada, J., P. N. Trathan, K. Reid, E. J. Murphy, and J. P. Croxall. 2006. Contrasting population changes in sympatric penguin species in association with climate warming. *Global Change Biology* 12(3): 411–423.

Fraser, W. R., W. Z. Trivelpiece, D. G. Ainley, and S. G. Trivelpiece. 1992. Increases in Antarctic penguin populations: Reduced competition with whales or a loss of sea ice due to environmental warming? *Polar Biology* 11(8): 525–531.

Lynnes, A. S., K. Reid, and J. P. Croxall. 2004. Diet and reproductive success of Adélie and chinstrap penguins: Linking response of predators to prey population dynamics. *Polar Biology* 27(9): 544–554.

Moline M. A., H. Claustre, T. K. Frazer, O. Schofield, and M. Vernet. 2004. Alteration of the food web along the Antarctic Peninsula in response to a regional warming trend. *Global Change Biology* 10(12): 1973–1980.

Newman, S. J., S. Nicol, D. Ritz, and H. Marchant. 1999. Susceptibility of Antarctic krill (*Euphausia superba* Dana) to ultraviolet radiation. *Polar Biology* 22(1): 50–55.

Nicol, S. 2006. Krill, currents, and sea ice: *Euphausia superba* and its changing environment. *BioScience* 56(2): 111–120.

Parkinson, C. L. 2004. Southern Ocean sea ice and its wider linkages: Insights revealed from models and observations. *Antarctic Science* 16(4): 387–400.

Turner, J., S. R. Colwell, and S. Harangozo. 1997. Variability of precipitation over the coastal western Antarctic Peninsula from synoptic observations. *Journal of Geophysical Research* 102(D12): 13999–14007.

Chapter 5
Now You "Sea" Ice, Now You Don't

Penguin communities shift on the
Antarctic Peninsula

Student Pages

Note: Reference List for Students

For more information on references cited in the Chapter 5 Student
Pages, go to teacher references on page 85.

STUDENT PAGE 5.1

Specialist Fact Sheet

Each Home Group contains five different specialists:

1. *Ornithologist:* A scientist who studies birds. Uses visual surveys (from ship or on land), diet analysis, bird banding, and satellite tracking to collect data on penguins.
2. *Oceanographer:* A scientist who studies the ocean. Uses satellite imagery, underwater sensors, and manual measurements of sea-ice thickness to collect data on sea-ice conditions and ocean temperature.
3. *Meteorologist:* A scientist who studies the weather. Uses automatic weather stations and visual observations of the skies to collect data on precipitation, temperature, and cloud cover.
4. *Marine Ecologist:* A scientist who studies relationships between organisms and their ocean environment. Uses visual surveys, diet analysis, and satellite tracking to collect data on a variety of organisms, including penguins.
5. *Fisheries Biologist:* A scientist who studies fish and their prey. Collects data on krill during research vessel cruises.

STUDENT PAGE 5.2

Ornithologists (Adélie Penguin Data Set)

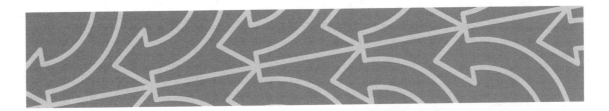

YEAR	# BREEDING PAIRS OF ADÉLIE PENGUINS
1975	15,202
1979	13,788
1983	13,515
1986	13,180
1987	10,150
1989	12,983
1990	11,554
1991	12,359
1992	12,055
1993	11,964
1994	11,052
1995	11,052
1996	9,228
1997	8,817
1998	8,315
1999	7,707
2000	7,160
2001	6,887
2002	4,059

Source: Data compiled from Smith, Fraser, and Stammerjohn. 2003. Photograph courtesy of Richard E. Lee, Jr.

- Adélie penguins spend their summers on land, where they breed. They spend winters on the outer extent of the sea ice surrounding Antarctica, where they molt their feathers and fatten up.
- Adélies are visual predators, meaning they need enough light to see their prey. Near the outer part of the pack ice, there are only a few hours of daylight in the middle of the winter. There is less sunlight as you go farther south (closer to land).
- On the western Antarctic Peninsula, Adélie penguins mostly eat krill, a shrimplike crustacean.
- Several countries have been heavily harvesting krill since the mid-1960s.
- Adélie penguins need dry, snow-free places to lay their eggs. They use the

same nest sites each year and at about the same time every year. Heavy snowfalls during the nesting season can bury adult Adélies and kill their eggs.

- Female Adélies lay two eggs, but usually only one of those eggs results in a fledged chick (fledged chicks have a good chance of maturing into adults). The two most common causes of death of eggs and chicks are abandonment by the parents (if they cannot find enough food) and predation by skuas (hawklike birds).

- In the water, Adélies are eaten mostly by leopard seals and killer whales.

- Adélies can look for food under sea ice because they can hold their breath for a long time. They are not as good at foraging in the open ocean, because they cannot swim very fast.

- Adélie penguins have lived in the western Antarctic Peninsula for at least 644 years.

STUDENT PAGE 5.3

Oceanographers (Sea Ice Data Set)

YEAR	AREA OF SEA ICE EXTENDING FROM THE ANTARCTIC PENINSULA (KM²)
1980	146,298
1981	136,511
1982	118,676
1983	88,229
1984	85,686
1985	78,792
1986	118,333
1987	142,480
1988	90,310
1989	44,082
1990	79,391
1991	111,959
1992	110,471
1993	94,374
1994	103,485
1995	95,544
1996	86,398
1997	100,784
1998	73,598
1999	79,223
2000	79,200
2001	69,914

Source: Data compiled from the Palmer Station, Antarctica Long-Term Ecological Research (LTER) data archive. Data from the Palmer LTER archive were supported by the Office of Polar Programs, NSF Grants OPP-9011927, OPP-9632763, and OPP-021782. Photograph courtesy of Marianne Kaput.

- In August or September (the middle of winter), sea ice covers over 19×10^6 km² of the Southern Ocean (an area larger than Europe). In February (the middle of summer), only 3×10^6 km² of the ocean is covered by sea ice.
- Sea ice keeps the air of the Antarctic region cool by reflecting most of the solar radiation back into space.
- Open water absorbs solar radiation instead of reflecting it and converts it to heat. This heat warms up the atmosphere.
- Sea ice reduces evaporation of the ocean, thus reducing the amount of moisture that is released to the atmosphere.
- As sea ice melts, bacteria and other particles are released into the

atmosphere. These particles can form condensation nuclei, which grow into rain or snow.

- Rain helps to stabilize the sea ice by freezing on the surface.
- Sea ice can be broken up by strong winds that last a week or more.

- An icebreaker is a ship with a reinforced bow to break up ice and keep channels open for navigation. Icebreakers were first used in the Antarctic in 1947 and have been commonly used to support scientific research for the last 25 years.

STUDENT PAGE 5.4

Meteorologists (Winter Snow Data Set)

YEAR	% OF PRECIPITATION EVENTS THAT ARE SNOW
1982	49
1983	67
1984	72
1985	67
1986	81
1987	80
1988	69
1989	69
1990	68
1991	72
1992	70
1993	70
1994	83
1995	77
1996	74
1997	81
1998	81
1999	83
2000	77
2001	90
2002	82
2003	76

Source: Data compiled from Antarctic Meteorology Online, British Antarctic Survey (www.antarctica.ac.uk/met/metlog). Photograph courtesy of Luke Sandro.

- In the winter, most of the precipitation in the western Antarctic Peninsula occurs as snow. There is an even mix of snow and rain the rest of the year.
- It is difficult to accurately measure the amount of snowfall in the Antarctic because strong winds blow the snow around.
- The Antarctic Peninsula has a relatively warm maritime climate so it gets more rain and snow than the rest of the Antarctic continent.
- Most of the rain and snow in the western Antarctic Peninsula is generated by cyclones from outside the Southern Ocean. Cyclones are areas of low atmospheric pressure and rotating winds.
- When there is less sea ice covering the ocean, there is more evaporation of the

ocean and, therefore, more moisture in the atmosphere.

- As sea ice melts, bacteria and other particles are released into the atmosphere. These particles can form condensation nuclei, which grow into rain or snow.

STUDENT PAGE 5.5

Marine Ecologists (Chinstrap Penguin Data Set)

YEAR	# BREEDING PAIRS OF CHINSTRAP PENGUINS
1976	10
1977	42
1983	100
1984	109
1985	150
1989	205
1990	223
1991	164
1992	180
1993	216
1994	205
1995	255
1996	234
1997	250
1998	186
1999	220
2000	325
2001	325
2002	250

Source: Data compiled from Smith, Fraser, and Stammerjohn. 2003. Photograph courtesy of Michael Elnitsky.

- Chinstrap penguins breed on land in the spring and summer and spend the rest of the year in open water north of the sea ice. The number of chinstraps that successfully breed is much lower in years when the sea ice does not melt until late in the spring.
- Chinstraps mostly eat krill, a shrimp-like crustacean.

- Whalers and sealers overhunted seals and whales, which also eat krill, until the late 1960s.
- Chinstraps hunt primarily in open water because they cannot hold their breath for very long.
- The main predators of chinstraps are skuas (hawklike birds), leopard seals, and killer whales.
- Chinstraps will aggressively displace Adélie penguins from nest sites in order to start their own nests and may compete with Adélies for feeding areas.
- Although chinstrap penguins have occupied the western Antarctic Peninsula for over 600 years, they have become numerous near Palmer Station (one of the three U.S. research stations in Antarctica) only in the last 35 years.

STUDENT PAGE 5.6

Fisheries Biologists (Krill Data Set)

YEAR	DENSITY OF KRILL IN THE SOUTHERN OCEAN (# KRILL/M²)
1982	91
1984	50
1985	41
1987	36
1988	57
1989	15
1990	8
1992	7
1993	22
1994	6
1995	9
1996	31
1997	53
1998	46
1999	4
2000	8
2001	31
2002	8
2003	3

- Several countries have been harvesting krill since the mid-1960s.
- Ultraviolet radiation is harmful to krill and can even kill them. Worldwide, ozone depletion is highest over Antarctica.
- Salps, which are small, marine animals that look like blobs of jelly, compete with krill for food resources. As the salt content of the ocean decreases, salps increase and the favorite food species of krill decrease.

Source: Data compiled from Atkinson et al. 2004. Photograph courtesy of Richard E. Lee, Jr.

- Krill, a shrimplike crustacean, is a keystone species, meaning it is one of the most important links in the Antarctic food web. All the vertebrate animals in the Antarctic either eat krill or another animal that eats krill.
- Krill eat mostly algae. In the winter, the only place algae can grow is on the underside of sea ice.

STUDENT PAGE 5.7

Specialist Group Report Sheet

Name: _____

Specialist Group: _____

In your own words, summarize the general trends or patterns of your data. Attach a graph of your data to the back of this sheet.

List possible explanations for the patterns you are seeing.

With the help of the facts on each data sheet, choose the explanation that you think is most likely. Why do you think that explanation is most likely?

Chapter 6
Population Peril

Polar bears decline in the
Canadian Arctic

Teacher Pages

AT A GLANCE

Rising air temperatures have changed the extent and timing of sea ice
formation in the Arctic, forcing some polar bear populations to go
longer each year without food. In this activity, students assume the
role of graduate students advising an intern participating in a polar
bear study. The students investigate declines in the body condition
and population size of polar bears, and reflect on the role of science in
wildlife management. (Class time: 1.5–3.5 hours)

"The Arctic is not a forsaken wasteland to a polar bear; it is home, and a comfortable home at that. For thousands of years, the climate, the ice, and the seals upon which it feeds have shaped and finely tuned the evolution of this predator so exquisitely that it has become not just a symbol but the very embodiment of life in the Arctic."

—Ian Stirling, *Polar Bears*, 1998

INTRODUCTION

The polar bear has become the poster child of Arctic climate change. The species is charismatic, instantly recognizable, and iconic of icy environments. Although polar bears are not in immediate danger of extinction, anecdotal and scientific evidence from around the Arctic suggest that climate change is bad news for bears (Aars, Lunn, and Derocher 2006). Evidence aside, the image of a polar bear clinging to a tiny piece of ice in a vast expanse of blue has emotional impact. Students exposed to news stories about the effects of climate change on polar bears may have strong preconceptions (or misconceptions) about bear ecology and the role of science in policy.

In this activity, students not only analyze data on polar bear populations, but also reflect on their personal opinions about wildlife conservation and the societal value of science. By the end of the lesson, students should be able to do the following:

- Explain how changes in habitat can affect the size and structure of wildlife populations.
- Describe some of the ways humans affect other species.
- Discuss the role of science in policy decisions about wildlife.
- Write a persuasive e-mail to a high school intern that is logically organized, scientifically accurate, and grammatically correct.

ICE IS LIFE

Polar bears (Figure 6.1) spend most of their time on annual sea ice near the coast of continents and island chains of the Arctic. They primarily hunt ringed seals (*Phoca hispid*) and bearded seals

(*Erignathus barbatus*), although they will infrequently kill whales and walruses. Both seal species use ice as a platform on which to rest, bear young, and shed their winter coat (Stirling 1998).

Although polar bears hunt all life stages of seals, almost or newly weaned pups are particularly important. Bears are at their lightest in March, just before seal pups are born. They are at their heaviest in midsummer, when the ice breaks up and pups disperse into open water. Ringed seals give birth in lairs excavated in snow drifts that

Figure 6.1

Polar bear (*Ursus maritimus*).
Polar bears are white to yellowish bears with a narrow head, small ears, and wide, stocky build. Adult females weigh 150–250 kg and males weigh 350–650 kg (Stirling 1998).
Polar bears are found in 19–20 subpopulations around the Arctic, nine of which are in Canada and four of which are shared between Canada and Alaska or Greenland. As of 2005, there were 20,000 to 25,000 bears worldwide (Aars, Lunn, and Derocher 2006).

Photograph courtesy of Lara Gibson.

form on the surface of the sea ice. When a polar bear detects a pup in its birth lair, it collapses the roof with its front feet and—if the pup hasn't escaped into the water—kills it with bites to the head and neck. Seals that are only a few months old have little experience with predators, which makes them relatively easy to catch. Seal pups are also high in calories. At six weeks old, 65–70% of a pup's weight is fat (Stirling 1998).

Some of the best-studied polar bears in the world spend their summers on the shores of western Hudson Bay, near Churchill, Manitoba (see Student Page 6.1, Figure 6.6, p. 113). Every summer, the ice on Hudson Bay completely melts (Figure 6.2, p. 102). Bears that have been hunting over the surface of the bay all winter and spring go ashore along the coast of Manitoba and Ontario, where the ice melts last. Other than the occasional snack of small birds or berries, the bears eat nothing in the summer (DeMaster and Stirling

1981). This means that all bears in the population live on their fat reserves for at least four months and pregnant females do so for eight months. Females give birth sometime between late November and early January, when other bears are already out on the newly formed ice. These mothers stay on land with their cubs until late February or early March (Stirling 1998).

The western coast of Hudson Bay is quite wet. There is little drainage because the land is flat and partially frozen the entire year. In the winter, northwesterly winds swooping over the icy cap of the bay bring arctic air and the occasional storm. In the summer, Hudson Bay is ice-free but still cold, and warm air traveling over the bay results in fog and rain. The area is characterized by tundra along the coast

Figure 6.2

Annual cycle of sea ice on Hudson Bay.

You also can find an animation of this cycle at *http://ice-glaces.ec.gc.ca/ wsvpageDsp.cfm?Lang=eng&ID=11860.*

Source: Figure from Stirling et al. 1977. © 1977, I. Stirling. Modified with permission from I. Stirling.

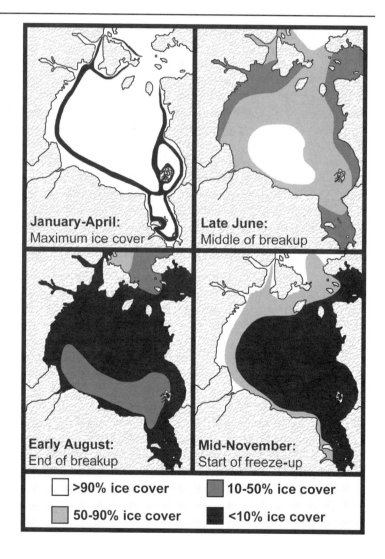

January–April: Maximum ice cover

Late June: Middle of breakup

Early August: End of breakup

Mid-November: Start of freeze-up

☐ >90% ice cover   ▨ 10–50% ice cover

▨ 50–90% ice cover   ■ <10% ice cover

(Figure 6.3) and peat bogs with small clumps of trees further inland (Stirling et al. 1977).

Bears spend over 90% of the ice-free season resting (Figure 6.4, p. 104). In part, they rest to avoid overheating. Polar bears are susceptible to heat stress because of their thick insulating layers of fur, skin, and blubber. Inactivity also helps bears save energy during the long months with no food (Stirling 1998). It's particularly important for female bears to save energy because the fatter a female is, the more

Figure 6.3

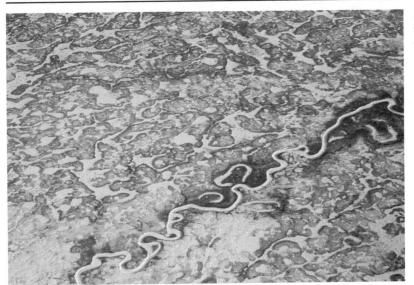

Aerial and ground-level views of the frozen, windswept coastline of western Hudson Bay.

Photographs courtesy of Richard E. Lee, Jr.

Figure 6.4

Polar bear sleeping in the tundra near Churchill.

Photograph courtesy of Lara Gibson.

likely her cubs will survive to weaning (Parks, Derocher, and Lunn 2006). Pregnant females may avoid heat and biting insects by entering maternity dens dug into permafrost banks as early as August (Richardson, Stirling, and Kochtubajda 2007). Despite these measures, adult bears can lose nearly 1 kg of body mass per day during the ice-free season (Derocher, Lunn, and Stirling 2004).

GOING, GOING, GONE?

Before an international agreement came into effect in 1976, overhunting was a serious threat to polar bears. Today, hunting is tightly controlled (or not allowed) in four of the five "polar bear nations." However, polar bears appear to face a new threat. Many experts believe that shrinking sea ice will cause a 30–60% reduction in the global polar bear population in the next 50 years (Aars, Lunn, and Derocher 2006; Amstrup, Marcot, and Douglas 2007). In May 2008, polar bears were listed as a threatened species in the United States because of changes in their sea ice habitat.

The ice-free season of Hudson Bay has become significantly longer since 1971, likely because of increases in air temperature and changes in atmospheric circulation (Gough, Cornwell, and Tsuji 2004). At the same time, the body condition and survival of polar bears has declined (Regehr et al. 2007; Stirling and Parkinson 2006). When the ice breaks up early, bears go without food for a longer period of time. Young bears, females with cubs, and old bears (more than 20 years old) are particularly vulnerable to starvation. In lean times, they may search for food in human settlements, which increases the risk they'll be shot. In the last few decades, conservation officers in Manitoba have handled a rising number of **problem bears**, and hunters in Nunavut report seeing bears where they have rarely— or never—been seen before (Stirling and Parkinson 2006). Furthermore, although cannibalism in polar bears is fairly rare, scientists have recently reported large, possibly undernourished males stalking and eating females and cubs (Amstrup et al. 2006).

Continuing changes in the Arctic will probably worsen the plight of polar bears. For example, as ice cover continues to decline, bears will be forced to swim more often. Although polar bears are excellent swimmers, it takes more energy for them to travel long distances in water than on ice, and they can drown under severe conditions (Monnett and Gleason 2006). Warm weather and increasing rain in the winter and spring also may negatively affect polar bears by collapsing their maternity dens. Females and their cubs can be crushed by their snow dens after periods of warm weather and high winds (Clarkson and Irish 1991). Warm and wet weather also can melt the birth lairs of ringed seals, exposing pups to extreme cold and increased risk of predation. Although bears have an easier time finding pups exposed on the ice, they also have increased competition from other predators such as Arctic foxes and gulls. Increased pup mortality could lead to a significant decline in populations of seals, and eventually, of polar bears (Stirling and Smith 2004).

Activity Overview

In this cross-curricular activity, students assume the role of graduate students mentoring a high school intern who is volunteering on a polar bear study near Churchill. The students work in small groups, and then individually, to answer questions posed by the intern in a series of e-mails. The e-mails are wide-ranging and discuss changes in polar bear habitat, the importance of bears to local communities, and perceived costs and benefits of climate change. Please note these e-mails are a fictional account of an amalgamation of research studies—not all of the studies were performed in the same year or by the same researcher.

TEACHING NOTES

Prior Knowledge

Although students can complete this activity with little or no prior knowledge, you may wish to have them do reading or research on the Arctic and/or polar bears. For example, students can learn more about polar bears on the Polar Bear International website *(www. polarbearsinternational.org)*. Click on the "Look at Polar Bears" link to find photographs and videos. You also may wish to show the "Pole to Pole" or "Ice World" episodes of Discovery's Planet Earth series (or a similar informative video) to pique student interest. Students new to climate change may have difficulty giving thorough answers to some of the discussion questions, so you might want to provide Chapter 4, "Quick Guide to Climate," as an extra resource.

Materials

- Student Page 6.1: Scenario and Polar Bear 101 (one per group)
- Student Pages 6.2–6.5: E-mails (one each per group)
- Optional: Computers with an internet connection or access to the library

Procedure: Class Discussions

1. Hand out Student Page 6.1. Introduce the lesson by reading the scenario aloud.
2. Split the class into small discussion groups (two to five students). Consider assigning some or all of the following roles to students within each group.
   a. *Reader*—reads each e-mail aloud to the group.
   b. *Recorder*—records the group's responses to the questions. The recorder may answer questions as a bulleted list rather than complete sentences.
   c. *Student representative*—interacts with the teacher if anyone in the group has a question.
   d. *Timekeeper*—monitors the time and keeps the group on task.
   e. *Reporter*—reports the group's responses to the rest of the class.
3. Hand out the first e-mail (Student Page 6.2). Give students 15–25 minutes to discuss it and to answer the questions. Depending on your curricular needs and time constraints, you may wish to have students answer only a subset of the questions.

4. At the end of the discussion period, call on just one or two groups to share their responses with the class. You may wish to keep track of responses on a blackboard or flipchart, either as a bulleted list or a concept map.

5. To give the illusion of time passing and to generate a bit of mystery, repeat Steps 3 and 4 with Student Pages 6.3–6.5 on three successive days. You could use the e-mail discussions as daily warm-ups or to wrap up a class. Note that depending on the level of your students, you may wish to allot additional time to the graphing activity in Student Page 6.4.

Procedure: Individual Assignments

Have each student write responses to two of the e-mails from the graduate student's point of view. This independent assignment may be completed in class or as homework. Students can choose which e-mails to respond to and can use details from their group discussions and/or additional research. You may wish to set a minimum number of words for the written responses.

Assessment

A grade for this lesson can be based on class discussions, written responses to e-mails, or both. You might want to address literacy standards by team teaching this lesson with a language arts instructor.

Because the writing assignment is in the format of an e-mail, students may slip into their normal "online" mode (i.e., lack of capitalization and/or punctuation, incomplete sentences, verbal shortcuts such as "LOL," etc.). Remind the students that in this scenario, they are helping someone in a professional capacity and so they should present themselves in a serious, respectful way. You may wish to provide a rubric to students before they complete the written assignment. Please fill in the "Points Possible" column before you photocopy the rubric (Figure 6.5, p. 108).

Figure 6.5

Rubric for written responses to questions.

STUDENT NAME:			
PERFORMANCE CRITERIA	**POINTS POSSIBLE**	**POINTS ACHIEVED**	**COMMENTS**
Responses include explanations for all aspects of the questions.			
Responses show an understanding of relationships among science facts and concepts.			
Responses show an understanding of relationships between scientific and social issues.			
Content and language of responses are relevant to the needs of the audience.			
Responses are organized logically and expressed clearly.			
Responses are written in complete sentences with no errors in grammar or spelling.			
TOTAL			

Source: Performance criteria adapted from Lantz 2004.

Modifications

This activity is ideal for making connections to academic standards in social studies and literacy. For example, the western shores of Hudson Bay have had a rich cultural history, being settled initially by **Inuit**, Dene, and Cree, and eventually by Europeans and Métis (Parks Canada 2005). Students could write plays, poetry, or essays about the importance of polar bears to various cultural groups of the North. Alternatively, students could investigate the potential effects of climate change on Inuit who still use traditional hunting practices. Finally, you may wish to have students explore polar bear management by the government of Manitoba, which gets revenue from ecotourism, and by the government of Nunavut, which gets revenue from hunting. Because polar bears born in Manitoba are eventually hunted in Nunavut, the two governments need to cooperate when managing bears (I. Stirling, personal communication, October 2007).

CONCLUSION

Climate change poses an imminent risk to polar bear populations. As with any other species, the management of polar bears will involve scientific, political, economic, and other societal components. In this activity, students examine polar bear declines and also consider why and how local and national governments should manage bears. More broadly, high school and first-year undergraduate students interested in life sciences are often unaware of career choices other than those in medicine; this activity may alert students to volunteer and career opportunities in wildlife biology.

REFERENCES

Aars, J., N. J. Lunn, and A. E. Derocher, eds. 2006. Polar bears: Proceedings of the 14th Working Meeting of the IUCN/SSC Polar Bear Specialist Group, 20–24 June 2005, Seattle, Washington, USA and Gland, Switzerland: IUCN.

Amstrup, S. C., B. G. Marcot, and D. C. Douglas. 2007. Forecasting the range-wide status of polar bears at selected times in the 21st century. U.S. Geological Survey, Reston, Virginia. Retrieved September 10, 2007, from *www.usgs.gov/newsroom/special/polar_bears*

Amstrup, S. C., I. Stirling, T. S. Smith, C. Perham, and G. W. Thiemann. 2006. Recent observations of intraspecific predation and cannibalism among polar bears in the southern Beaufort Sea. *Polar Biology* 29(11): 997–1002.

Clarkson, P. L., and D. Irish. 1991. Den collapse kills female polar bear and two newborn cubs. *Arctic* 44(1): 83–84.

DeMaster, D. P., and I. Stirling. 1981. *Ursus maritimus. Mammalian Species* 145: 1–7.

Derocher, A. E., N. J. Lunn, and I. Stirling. 2004. Polar bears in a warming climate. *Integrative and Comparative Biology* 44(2): 163–176.

Gough, W. A., A. R. Cornwell, and L. J. S. Tsuji. 2004. Trends in seasonal sea ice duration in southwestern Hudson Bay. *Arctic* 57(3): 299–305.

Lantz, H. B. 2004. *Rubrics for assessing student achievement in science grades K–12.* Thousand Oaks, CA: Corwin Press.

Monnett, C., and J. S. Gleason. 2006. Observations of mortality associated with extended open-water swimming by polar bears in the Alaskan Beaufort Sea. *Polar Biology* 29(8): 681–687.

Parks, E. K., A. E. Derocher, and N. J. Lunn. 2006. Seasonal and annual movement patterns of polar bears on the sea ice of Hudson Bay. *Canadian Journal of Zoology* 84(9): 1281–1294.

Parks Canada. 2005. Wapusk National Park of Canada: Natural

wonders and cultural treasures. Retrieved September 9, 2007, from *www.pc.gc.ca/pn-np/mb/wapusk/natcul/natcul1a_e.asp*

Regehr, E.V., N. J. Lunn, S. C. Amstrup, and I. Stirling. 2007. Effects of earlier sea ice breakup on survival and population size of polar bears in western Hudson Bay. *Journal of Wildlife Management* 71(8): 2673–2683.

Richardson, E., I. Stirling, and B. Kochtubajda. 2007. The effects of forest fires on polar bear maternity denning habitat in western Hudson Bay. *Polar Biology* 30(3): 369–378.

Stirling, I. 1998. *Polar bears.* Ann Arbor, MI: University of Michigan Press.

Stirling, I., C. Jonkel, P. Smith, R. Robertson, and D. Cross. 1977. The ecology of the polar bear (*Ursus maritimus*) along the western coast of Hudson Bay. Canadian Wildlife Service, Occasional Paper No. 33.

Stirling, I., and C. L. Parkinson. 2006. Possible effects of climate warming on selected populations of polar bears (*Ursus maritimus*) in the Canadian Arctic. *Arctic* 59(3): 261–275.

Stirling, I., and T. G. Smith. 2004. Implications of warm temperatures and an unusual rain event for the survival of ringed seals on the coast of southeastern Baffin Island. *Arctic* 57(1): 59–67.

Wilmers, C. C., and W. M. Getz. 2005. Gray wolves as climate change buffers in Yellowstone. *PLoS Biology* 3(4): 0571–0576. Published in 2005 by the Library of Science.

OTHER RECOMMENDED RESOURCES

These additional resources were used to create the Student Pages, but are not cited in the text:

Canadian Wildlife Service. 2002. Field projects: Polar bears. Retrieved August 17, 2007, from *www.mb.ec.gc.ca/nature/ecb/da02s14.en.html*

Indian and Northern Affairs Canada. 2004. Words first: An evolving terminology relating to Aboriginal peoples in Canada. Retrieved September 6, 2007, from *www.ainc-inac.gc.ca/pr/pub/wf/index_e.html*

Ferguson, S. H., I. Stirling, and P. McLoughlin. 2005. Climate change and ringed seal (*Phoca hispida*) recruitment in western Hudson Bay. *Marine Mammal Science* 21(1): 121–135.

Fischbach, A. S., S. C. Amstrup, and D. C. Douglas. 2007. Landward and eastward shift of Alaskan polar bear denning associated with recent sea ice changes. *Polar Biology* 30(11): 1395–1405.

Freeman, M. M. R., and G. W. Wenzel. 2006. The nature and significance of polar bear conservation hunting in the Canadian Arctic. *Arctic* 59(1): 21–30.

Manitoba Conservation. n.d. Great white bears of Manitoba. Retrieved September 10, 2007, from *www.gov.mb.ca/conservation/wildlife/managing/polar_bears/index.html*

Chapter 6
Population Peril

Polar bears decline in the
Canadian Arctic

Student Pages

Note: Reference List for Students

For more information on references cited in the Chapter 6 Student
Pages, go to teacher references on page 109.

STUDENT PAGE 6.1

Scenario and Polar Bear 101

Scenario

You are a graduate student at a local university who is studying bears. Over the past year, you have visited a number of high schools to talk about your work and to encourage students to get involved in science. One day, a senior named Christy Martinez sends you an e-mail. She is going to the Canadian Arctic, west of Hudson Bay, to study polar bears and wants to know if it's OK to contact you with any questions she might have. Although you're a bit jealous of Christy's opportunity (you've always wanted to see a polar bear in the wild), you're flattered that she wants your help. You look forward to getting her first e-mail from the Arctic.

Polar Bear 101

- Polar bears live in Canada, the United States (Alaska), Russia, Norway, and Greenland (which is part of Denmark). There are 20,000 to 25,000 bears worldwide.
- Polar bears eat mostly ringed seals and, to a lesser degree, bearded seals. The bears must hunt seals on or near sea ice because they can't catch them in open water.
- Hudson Bay (Figure 6.6) is covered by sea ice up to eight months of the year. When the ice melts in July, bears move onto land and survive on stored fat until early November.
- Near Churchill, Manitoba, polar bears give birth in maternity dens they have dug in frozen peat. The dens may be reused for several years.
- Most polar bears spend the winter hunting on the sea ice, rather than hibernating in a den like grizzlies or black bears. In the summertime, females will sometimes rest in maternity dens to escape hot weather and biting insects.
- Polar bears are important to **Inuit**, the native people of the Canadian Arctic. (The word *Eskimo* is considered derogatory and is no longer used in Canada.) Inuit hunters can get $500 to $1000 (Canadian currency) for the hide of a polar bear. An average-size bear yields 140 kg of edible meat that is distributed to the community as food for people and dogs.

Figure 6.6

Map of Hudson Bay, Canada.

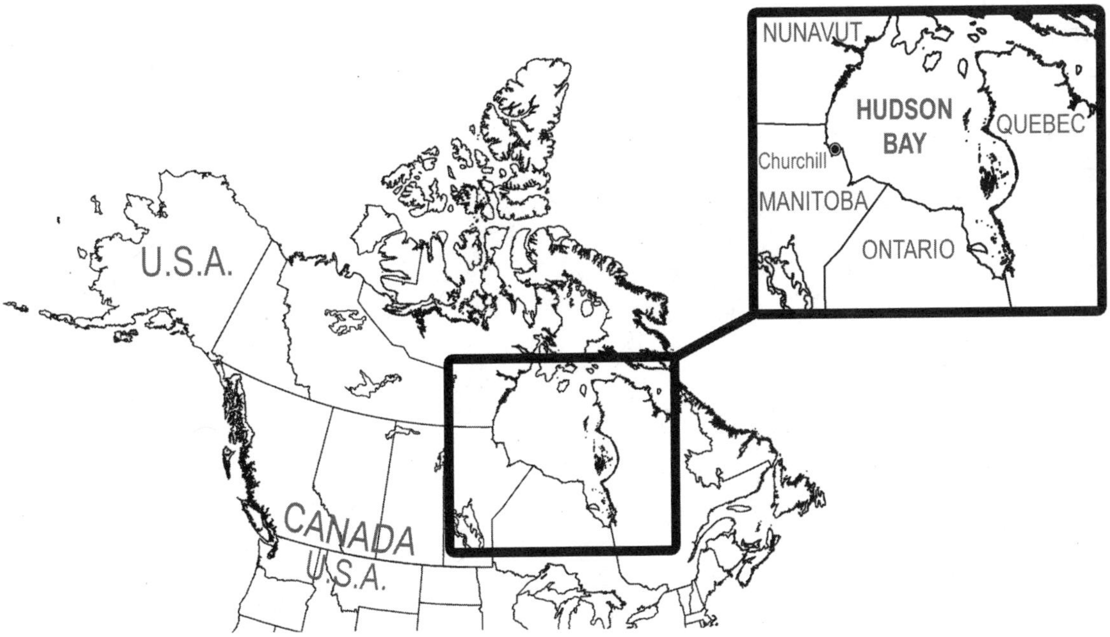

STUDENT PAGE 6.2

Arriving in Churchill

From: Martinez, Christy
Sent: August 8, 11:56 a.m.
Subject: Orientation

Hi:

I finally made it! There are no roads to Churchill, so I had to take a train from Winnipeg, Manitoba. I might look into flying back home, because the train took more than 36 hours.

The weather hasn't been very good so far. Even though it's August, it's only supposed to get to 12°C today and will cool off to 6°C tonight. It seems like we could use a little more global warming here....

I expected the landscape to be flat, but I didn't expect it to be so wet. There are little streams, ponds, and puddles of water everywhere. There are hardly any trees, and the few I do see are all tiny.

This morning, I went to an orientation for new volunteers. We were taught, "A safe bear is a distant bear," and that we can't leave the station on foot unless someone in the group is carrying a gun. Apparently the number of bear sightings around the station has increased (see the attached graph). I wanted to ask "Doc"—my new boss—about why this is happening, but we ran out of time. Can you help?

The lunch bell is ringing, so I'd better go.
Christy

Attachments:

Trend in number of bears handled by wildlife conservation officers in Churchill, Manitoba.

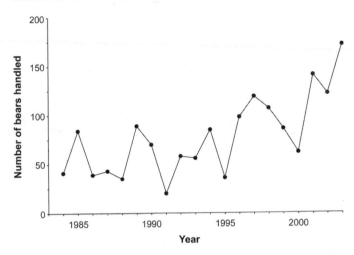

Source: Stirling, I., and C. L. Parkinson. 2006. © 2006. Arctic Institute of North America. Reproduced with permission from I. Stirling and the Arctic Institute of North America.

Polar Bear Alert

Photograph courtesy of Lara Gibson.

ANSWER THESE QUESTIONS:

1. Why do you think there are no permanent roads in and out of Churchill?

2. Why do you think the trees around Churchill are so small and few in number? (Hint: The answers to Questions 1 and 2 are related.)

3. When Christy got to Churchill, the weather was a bit cool (it's common for August days to be warmer than 20°C). Christy jokingly comments about the need for climate change in Churchill. Would she be able to tell from just a few days that climate change is occurring near Churchill? List at least two possible pros and cons of climate change in the Arctic.

4. Brainstorm some reasons for the increase in **problem bears**. (Problem bears are animals that have become accustomed to people and so are not afraid of them. They can threaten human life and property.)

5. What, if anything, do you think the government of Churchill should do about problem bears? Why? Consider the potential loss of human property and/or lives, costs to taxpayers, revenue from "bear tourism," and the increasing worldwide interest in polar bears in your answer.

STUDENT PAGE 6.3

Effect of Fire on Polar Bear Dens

From: Martinez, Christy
Sent: September 10, 8:02 p.m.
Subject: Re: Messy Roommates

Hey:

Thanks for your suggestions on how to deal with my roommates. It's hard sharing a room with seven other people.

For the last few weeks, I've been helping with a survey of polar bear dens in Wapusk National Park, which is southeast of Churchill. Once you get farther inland, the vegetation changes a lot. There are berry bushes and lots more trees.

Bears dig their dens in the side of permafrost banks (ground that stays frozen all year). Doc thinks that forest fires might increase in the future because of climate change and that fires might affect polar bear dens. So, we are comparing the physical structure of burned dens with the structure of unburned dens.

I've been making notes in my journal every night. Here's a summary of our results so far:

DEN CHARACTERISTICS	BURNED DENS	UNBURNED DENS
Thickness of unfrozen ground on the top of the dens	Higher	Lower
Stability of dens (firmness of the den walls)	Lower	Higher
Percent tree cover around dens	Lower	Higher
Number of dens with collapsed roof	Higher	Lower

But so what? And how likely is it that fires are going to be worse in the future anyway?

Sorry—I'm a bit cranky. You'd think frozen ground would be easy to walk on, but when there's an unfrozen layer on top, your feet sink like you're on soft snow. And then, the bushes snag at your boot laces and the bottom of your pants.... I think I spent more time on my rear than on my feet today!

Later,
Christy

ANSWER THESE QUESTIONS:

1. Most fires in the Hudson Bay area start because of a lightning strike. Why do you think the number of fires in that area might increase as the climate gets warmer?

2. Study the table Christy sent you. What is the effect of forest fires on polar bear dens? What direct effects might this have on polar bears? (Hint: Remember that polar bears go a long time without eating in the summer.)

STUDENT PAGE 6.4

Capturing and Collaring Polar Bears

From: Martinez, Christy
Sent: September 23, 2:12 p.m.
Subject: Helicopter Ride

I got to help capture a polar bear today!

We used a helicopter to search for bears—it was easy to spot the white bears on the dark background of the land. The weather was cold, though. Doc tries to capture bears only in cool weather because they overheat easily if they get stressed.

Doc shot a tranquillizer dart at an adult female from the helicopter. After the bear stumbled and lay down, the pilot landed and we approached her on foot. It took about 10 minutes for her to be completely asleep.

At first, when we were standing next to the bear and I saw how big she was, I was so nervous that I couldn't understand what Doc wanted me to do. I calmed down after a couple of minutes and helped measure her body length (from the tip of her nose to the tip of her tail) and the circumference around her chest. Later, we can calculate her weight by plugging length and circumference into an equation. Doc was really picky about my technique—he made me practice again and again until I got the same measurement each time.

We also put a GPS collar on the bear. Every few days, a satellite will collect data on the bear's location and the surrounding temperature. Apparently we can't put collars on males, because their heads are smaller than their necks and the collars slip right off. Do you think that might bias our study results?

We hiked back to the helicopter, keeping one eye on the bear and another on some dark clouds. Thankfully, the bear fully recovered from the anesthetic and we were able to take off before a big storm hit.

We just got back to the station, and it looks like we'll be inside for the rest of the day because of the weather. Doc gave me two data tables (see attachments) and wants me to make a graph showing body mass and the date each year that the sea ice breaks up. Help! I don't know the best way to do that.

Thanks,
Christy

Attachments:

Trend in body mass of adult female polar bears in western Hudson Bay, Canada.

Bears were captured in the autumn, near the end of the ice-free season.

YEAR	MEAN MASS OF BEARS (KG)
1980	296.2
1981	287.1
1982	368.6
1983	252.9
1984	278.6
1985	240.7
1986	278.6
1987	275
1988	258.8
1989	271.3
1990	261
1991	247.1
1992	273.5
1993	275
1994	244.8
1995	245.6
1996	252.9
1997	208.8
1998	255.1
1999	237.5
2000	287.1
2001	240.7
2002	205.7
2003	230.1
2004	237.5

Source: Data compiled from Stirling and Parkinson 2006.

Trend in sea ice breakup date in western Hudson Bay, Canada.

The breakup date is the point at which sea ice covers <50% of the ocean in a given area.

YEAR	BREAKUP DATE
1980	June 28
1981	June 26
1982	July 5
1983	July 7
1984	June 29
1985	July 4
1986	June 25
1987	July 4
1988	July 1
1989	June 17
1990	June 7
1991	June 22
1992	July 10
1993	June 25
1994	June 19
1995	June 13
1996	June 20
1997	June 15
1998	June 9
1999	June 10
2000	June 25
2001	June 15
2002	June 23
2003	May 25
2004	July 2

Source: Data compiled from Stirling and Parkinson 2006.

ANSWER THESE QUESTIONS:

1. Scientists can only use collars with global positioning systems (GPS) to study the movement patterns and habitat choices of female polar bears. Do you think males and females might use their environment differently? Why or why not? Suggest two ways scientists with unlimited time and money could collect similar data on male bears.

2. Why do you think Christy's boss is interested in knowing the relationship between body mass and sea ice? (Hint: Why is sea ice important to polar bears?)

3. Describe the best way to combine these data sets into one graph. (You can plot the data yourself, if it will help you, but don't show Christy the graph.) Tell Christy which variable to put on which axis and why. Remind her of the elements of a good graph (a clear title, labels on each axis that indicate the unit of measurement, appropriate scale, and the entire range of data).

4. Why does the body mass data only include adult females with no cubs? In other words, why didn't the scientists also include adult males, young bears, or females with cubs?

STUDENT PAGE 6.5

Extinction?

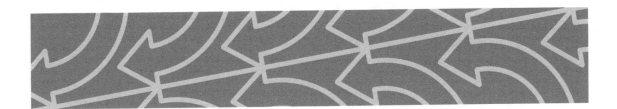

From: Martinez, Christy
Sent: November 5, 1:27 a.m.
Subject: Leaving Tomorrow

Every muscle in my body hurts. We've been working long days, trying to finish up last-minute research before the end of the field season. Hudson Bay is starting to freeze up, so the bears will move out onto the ice soon. I just finished packing my stuff for the trip home. I should go to bed, but I don't think I can sleep just yet.

Yesterday, all the scientists at the research station gave presentations about their work. Most of the talks had something to do with the warming climate—how it's affecting plants, snow geese, and even people. Doc presented current knowledge on the effects of climate change on bears. Here are some new things I learned:

- In the western Hudson Bay area over the last 20–30 years,
  - the air temperature has increased.
  - the sea ice has broken up earlier in the year.
  - fewer ringed seal pups have survived to adulthood.
  - on average, female polar bears weigh less in years that the ice breaks up early because they have to come to shore earlier. Females that weigh less than 189 kg can't reproduce.
  - the bear population has declined by 22% (from 1,194 bears in 1987 to 935 bears in 2004).
- Scientists who study other polar bear populations have reported the following:
  - The number of polar bears seen swimming in September at least 2 km from shore has increased. In 2004, four bears drowned near Alaska during rough ocean conditions. Scientists think the actual number of deaths was probably higher (floating bears are hard to detect from an airplane).
  - The number of polar bears giving birth on ice has decreased because the ice is becoming less stable. (Polar bears in northern Alaska can give birth either in earth dens on the coast or snow dens on the ice.)
  - Skinny male polar bears are killing and eating bear cubs.

I'm worried polar bears might go extinct because of the warming climate. And I don't know if there's anything we can do to save them....

Well, I'd better go. Talk to you soon.
Christy

ANSWER THESE QUESTIONS:

1. Based on the information presented here and in previous e-mails, do you think Christy's worries are justified? Consider how the survival of a species can be affected by changes to the species' food source, the quality of the species' habitat, and interactions with other species (including humans). Consider also how species can adjust to changes in their habitat or food sources.

2. What other data on polar bears (including from other areas in the Arctic) might be useful to determine if they are being affected by climate change?

3. Why are there laws protecting wildlife? What role do you think science plays in making these laws?

4. What, if anything, should be done to protect polar bears? Keep in mind they live in five different countries, each of which has a different policy on climate change, hunting, and wildlife management.

Chapter 7
Carrion: It's What's for Dinner

Wolves reduce the impact of climate change

Teacher Pages

AT A GLANCE

The restoration of wolves to Yellowstone National Park after a 70-year absence created a natural experiment on the ecological effects of top predators. In this activity, students use mathematical models to explore how carrion from wolf kills can reduce negative effects of climate change on scavengers in the park. (Class time: 1–2.5 hours)

This chapter is modified from Constible, J., L. Sandro, and R. Lee. 2008. Carrion—It's what's for dinner: Wolves reduce the impact of climate change. *American Biology Teacher* 70(2): 95–102.

"Who would not give a year of his life to see a wild wolf or a whole pack of wolves trailing down an elk or deer?"
—Edmund Heller, "The Big Game Animals of Yellowstone National Park," *Roosevelt Wildlife Bulletin*, 1925

"Mathematics is the science which uses easy words for hard ideas."
—Edward Kasner and James Newman, *Mathematics and the Imagination*, 1940

INTRODUCTION

Humans have viewed wolves as competitors, threats to personal safety, and symbols of evil throughout history. By the early part of the 20th century, gray wolves (*Canis lupus*) had been eradicated from 42% of their historic range in North America (Laliberte and Ripple 2004). In Yellowstone National Park, gray wolves were hunted to local extinction by 1926 but were reintroduced in 1995 after a decades-long process involving biologists, politicians, ranchers, and the general public (Table 7.1). By the end of 2006, the wolf population in the park was at least 136 wolves in 13 packs (Smith et al. 2007).

Table 7.1

Brief history of gray wolves in Yellowstone National Park.

1872	Yellowstone National Park (YNP) is established by an act of the U.S. Congress.
1872–1917	Wolves in and around YNP are killed for pelts, to protect humans and livestock, and for sport.
1918	The newly formed National Park Service takes control of YNP and continues to hunt wolves.
1926	Wolves are exterminated from YNP.
1973	The U.S. government lists the gray wolf as an endangered species.
1995–1996	Thirty-one wolves are reintroduced into YNP.

Source: Data compiled from Berger and Smith 2005; Phillips and Smith 1996.

By the end of the lesson, students should be able to do the following:
- Define and give examples of keystone species.
- Demonstrate, using mathematical models, that ecosystems are more resilient to environmental change when they contain a full complement of species, including top carnivores.
- Recognize that math is a vital tool in scientific investigations.

WOLVES AS KEYSTONES

From an ecological perspective, it was important to restore the gray wolf to Yellowstone because it is a **keystone species**. Keystone species, which are usually top predators, affect their communities or ecosystems in a much larger way than expected based on abundance alone (Steneck 2005). The presence, abundance, and productivity of a wide array of species in Yellowstone National Park are indirectly affected by interactions of wolves with elk and coyotes (Figure 7.1).

Figure 7.1

The wolf is a keystone species in Yellowstone National Park.

For example, the reintroduction of wolves has facilitated the recovery of beavers in Yellowstone. In the 1800s, human trappers decimated beaver populations. After wolves were removed from the park, elk populations grew and competition for willow—the preferred food and construction material of beavers—became intense (Ripple and Beschta 2003). Although beavers had been protected from trapping since the early 1920s and reintroduction efforts were underway, competitive pressure from elk suppressed beaver recovery (Baker et al. 2005). Since wolves were restored to Yellowstone, predation, hunting, and drought have reduced elk populations. But elk also have changed their behavior (Creel et al. 2005): When elk detect wolves in a general area, they move toward conifer forests (where they have good protection from wolves) and away from open areas and streams (where they have less protection from wolves) (Figure 7.2, p. 126). Because of

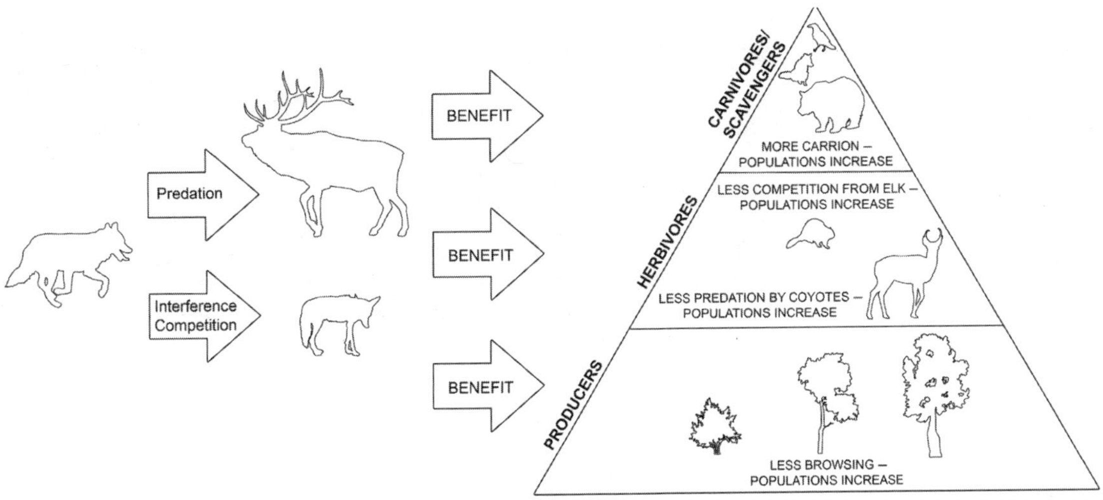

SC*LINKS*®
THE WORLD'S A CLICK AWAY

Topic: Wolves
Go to: *www.scilinks.org*
Code: CCPP13

the combined effect of fewer elk and reduced use of willow habitat, the number of beaver colonies (five to six beavers per colony) on the northern range of Yellowstone increased from one in 1996 to nine in 2003 (D. Smith, personal communication, October 2006).

Figure 7.2

An elk forages in a stream during winter.

Photograph courtesy of Ed Thomas.

WINTER ON THE NORTHERN RANGE

In Yellowstone National Park, daytime winter temperatures range from –20°C to –5°C and snow can exceed 7 m at high elevations (NPS 2006). Every autumn, the northern Yellowstone elk herd migrates from high-elevation summer habitat within the park to milder habitat in the northern range, a 1,530 km^2-area that includes a portion of the park and some adjacent public and private land (Singer and Mack 1999). Even so, winter isn't easy. Elk search for grasses and other herbaceous plants by digging in the snow with their hooves. When the snow is deep or covered by a hard crust, digging becomes more difficult, as does the simple act of moving through the snow (Gese, Ruff, and Crabtree 1996). Furthermore, plants under the snow are less nutritious than in the summer. In severe winters, elk regularly starve to death (Wisdom and Cook 2000). They also can suffer the same fate during mild winters if conditions during the previous summer were poor (Vucetich, Smith, and Stahler 2005).

Carcasses—particularly those of elk—are an important food source for Yellowstone's carnivores. Many carnivores such as bears and eagles scavenge carrion during winter and early spring (Figures 7.1 and 7.3). Some species, such as ravens, have even learned to track wolves to kill sites (Stahler, Heinrich, and Smith 2002). Before wolves were restored to Yellowstone, carrion availability depended on winter severity. In winters with deep snow and low temperatures, elk carrion was plentiful; in mild winters, carrion was sparse. During the rest of the year, carrion was negligible (Gese, Ruff, and Crabtree 1996). Even in the presence of wolves, snow cover plays a role in the amount of carrion in the park: Wolves leave more carrion for scavengers when snow is deep, because elk are easier to kill and wolf packs eat a smaller proportion of each kill. However, the presence of wolves also has altered the *timing* of carrion in the park. Carrion is now available year-round, regardless of the snow cover, and is a more predictable resource for scavengers (Wilmers et al. 2003a). The change in the timing and predictability of carrion benefits both small scavengers (e.g., foxes), which have small stores of body fat and need to feed frequently, and large scavengers (e.g., bears), which require a high-energy food source before hibernation (Wilmers and Getz 2005).

No other carnivores in Yellowstone fill the ecological role of the gray wolf. Coyotes occasionally kill elk, but they feed primarily on small mammals and carrion (Crabtree and Sheldon 1999). Bears prey on elk only during some parts of the year. Cougars are a major year-

Figure 7.3

After wolves killed this bison and had eaten their fill, the carcass was picked clean by scavengers.

Note how the snow around the carcass (foreground) has been churned up by a large number of animals.

Photograph courtesy of Ed Thomas.

round predator of elk, but they defend their kills from scavengers more fiercely than wolves and hide uneaten prey (Berger and Smith 2005). Finally, human hunters provide large amounts of carrion in the form of gut piles on park borders, but only from early January to mid-February. Bears in hibernation cannot take advantage of midwinter gut piles, and scavenging coyotes have difficulty finding the gut piles and are often shot by human hunters (Wilmers et al. 2003b).

LET IT SNOW!

As global temperatures rise, winter precipitation will fall as rain more often than as snow, and snowmelt will occur earlier in the spring (Barnett, Adam, and Lettenmaier 2005). Since 1948 winter temperatures have increased, the monthly snow depth has decreased, and the snow season has gotten shorter in the northern part of Yellowstone. Using mathematical models, two scientists in California demonstrated that although less carrion is available to scavengers as snow cover declines, the reduction is less dramatic when wolves are present in the park (Wilmers and Getz 2005). In essence, wolves act as a "buffer" against climate change by providing more carrion: They delay the detrimental effects of declining snow cover such that other species have more time to adapt to their changing environment. The presence of wolves might be especially important to threatened species such as grizzly bears. Climate change and disease have reduced the availability of white-bark pine (*Pinus albicaulis*), one of the few high-quality food sources available to Yellowstone bears in autumn. Wolf-killed elk may give bears time to adapt to a new food source (Smith and Ferguson 2005).

TEACHING NOTES

Materials

- Calculators or computers
- Pencils or pens
- Student Pages (1 copy per student)

Procedure

This activity may be done individually, in small groups (two to four students), or as a class.

1. Before class starts, write this question on the board: "Why are carnivores important in ecosystems?"
2. As students enter the room, engage their attention by playing a sound track of wolves howling and/or a video or slide show of Yellowstone carnivores/scavengers hunting or eating prey (see Table 7.2 for resources).
3. Once students are seated, draw attention to the question on the board and explain that today's class will focus on gray wolves in Yellowstone National Park. Introduce the term *keystone species.*
4. Introduce the park with a short slide show or video (see Table 7.2 for resources). Include the history of wolves in the park (Table 7.1 on p. 124).
5. Have students read the student pages and answer the questions associated with each graph and table. You may wish to circulate around the class to ensure that students are answering the questions thoughtfully.

Table 7.2

Internet resources.

- Red fox feeding on a moose carcass: www.admin.mtu.edu/urel/breaking/2006/Videos/redfox.mov
- Wolves feeding on an elk carcass: www.nps.gov/archive/yell/tours/thismonth/nov2004/wolves/index.htm
- Electronic field trip of Yellowstone National Park: www.windowsintowonderland.org/orientation/pages/index.html
- Video of park ranger discussing wolves: www.windowsintowonderland.org/wolves2/teacherinfo.shtml
- Yellowstone Park Digital Slide File: www.nps.gov/archive/yell/slidefile/index.htm

Assessment

The correct answers for the student worksheets are: Table 7.3—301 kg, 196 kg, and 91 kg; Table 7.5—1,275 kg, 1,239 kg, and 1,221 kg. To encourage critical thinking about what the numbers actually mean, however, we suggest providing a rubric to students at the start of the activity that includes performance criteria on data interpretation and logical scientific explanations (Figure 7.4, p. 130).

Figure 7.4

Assessment rubric.

Date:

Student Name(s):

PERFORMANCE CRITERIA	POINTS POSSIBLE	SELF	TEACHER
My mathematical calculations are accurate.	5		
I described all trends, patterns, and relationships shown by each graph and table.	10		
I made appropriate inferences and/or conclusions based on prior knowledge, background information, and the available data.	15		
I expressed my ideas clearly and logically.	15		
My spelling and grammar are correct.	5		
TOTAL	50		
NOTES:			

Source: Performance criteria adapted from Lantz 2004. Photograph courtesy of Yellowstone Wolf Project.

Modifications

An alternative way to introduce Yellowstone wolves to students is WolfQuest, a free video game available at *www.wolfquest.org*. In the game, students hunt elk, harass coyotes, and feed on carcasses dotted across a mountainous landscape.

You may wish to extend this lesson by having your students consider another ecosystem in which top carnivores can ameliorate the effects of climate change: Since 1979 there have been dozens of massive bleaching events, in which reef-forming corals lost their symbiotic algae in response to high temperatures. Recovery, while possible, is impeded by other environmental threats, such as predation by the crown-of-thorns starfish. Populations of starfish are kept in check partly by three commercially prized predators, including the triton snail. Make some predictions about how climate change would affect coral reefs in the presence and absence of starfish predators. You can get more information from these websites:

- AIMS Research—Coral Bleaching: *www.aims.gov.au/pages/research/coral-bleaching/coral-bleaching.html*
- CoRIS—Hazards to Coral Reefs: *www.coris.noaa.gov/about/hazards*
- CRC Reef Research Center: *www.reef.crc.org.au/discover/plantsanimals/cots/cotstheory.html*
- PBS Coral Reef Connections: *www.pbs.org/wgbh/evolution/survival/coral/predators.html*

As an alternative, have your students explore the social and political ramifications of wolf reintroduction by imagining what each of the following stakeholders might have to say about wolf reintroduction:

- A rancher, who raises sheep on the border of the park.
- An environmentalist, who wants to preserve the natural environment of the park.
- A tourist, who has come to the park to escape city life and to see wildlife.
- A scientist, who studies interactions between predators and their prey.
- A hunter, who has spent a large sum of money for a guided elk hunt north of the park.

CONCLUSION

This activity fits multiple aspects of the high school life science curriculum. The lesson addresses science education standards on populations, ecosystems, and environmental change, and students learn that math is a useful tool for answering complicated questions about the natural world. Finally, climate change lessons are often, by nature, full of doom and gloom. This activity suggests to learners that humans can mitigate the effects of a warming world by protecting the integrity and species diversity of ecosystems.

REFERENCES

Baker, B. W., H. C. Ducharme, D. C. S. Mitchell, T. R. Stanley, and H. R. Peinetti. 2005. Interaction of beaver and elk herbivory reduces standing crop of willow. *Ecological Applications* 15(1): 110–118.

Barnett, T. P., J. C. Adam, and D. P. Lettenmaier. 2005. Potential impacts of a warming climate on water availability in snow-dominated regions. *Nature* 438(7066): 303–309.

Berger, J., and D. W. Smith. 2005. Restoring functionality in Yellowstone with recovering carnivores: Gains and uncertainties. In *Large carnivores and the conservation of biodiversity*, eds. J. C. Ray, K. H. Redford, R. S. Steneck, and J. Berger, 100–109. Washington, DC: Island Press.

Crabtree, R. L., and J. W. Sheldon. 1999. The ecological role of coyotes on Yellowstone's northern range. *Yellowstone Science* 7(2): 15–23.

Creel, S., J. Winnie, B. Maxwell, K. Hamlin, and M. Creel. 2005. Elk alter habitat selection as an antipredator response to wolves. *Ecology* 86(12): 3387–3397.

Gese, E. M., R. L. Ruff, and R. L. Crabtree. 1996. Foraging ecology of coyotes (*Canis latrans*): The influence of extrinsic factors and a dominance hierarchy. *Canadian Journal of Zoology* 74(5): 769–783.

Laliberte, A. S., and W. J. Ripple. 2004. Range contractions of North American carnivores and ungulates. *BioScience* 54(2): 123–138.

Lantz, H. B. 2004. *Rubrics for assessing student achievement in science grades K–12*. Thousand Oaks, CA: Corwin Press.

National Park Service (NPS). 2006. Yellowstone weather. Retrieved September 5, 2007, from *www.nps.gov/yell/planyourvisit/weather. htm*

Phillips, M. K., and D. W. Smith. 1996. *The wolves of Yellowstone*. Stillwater, MN: Voyageur Press.

Ripple, W. J., and R. L. Beschta. 2003. Wolf reintroduction, predation risk, and cottonwood recovery in Yellowstone National Park. *Forest Ecology and Management* 184(1–3): 299–313.

Singer, F. J. and J. A. Mack. 1999. Predicting the effects of wildfire and carnivore predation on ungulates. In *Carnivores in ecosystems: The Yellowstone experience*, eds. T. W. Clark, A. P. Curlee, S. C. Minta, and P. M. Kareiva, 189–237. New Haven, CT: Yale University Press.

Smith, D. W., and G. Ferguson. 2005. *Decade of the wolf: Returning the wild to Yellowstone*. Guilford, CT: The Lyons Press.

Smith, D. W., D. R. Stahler, D. S. Guernsey, M. Metz, A. Nelson, E. Albers, and R. McIntyre. 2007. *Yellowstone Wolf Project: Annual report, 2006*. National Park Service, YCR-2007-01. Yellowstone National Park, WY: Yellowstone Center for Resources.

Stahler, D., B. Heinrich, and D. Smith. 2002. Common ravens,

Corvus corax, preferentially associate with grey wolves, *Canis lupus*, as a foraging strategy in winter. *Animal Behaviour* 64(2): 283–290.

Steneck, R. S. 2005. An ecological context for the role of large carnivores in conserving biodiversity. In *Large carnivores and the conservation of biodiversity*, eds. J. C. Ray, K. H. Redford, R. S. Steneck, and J. Berger, 9–33. Washington, DC: Island Press.

Vucetich, J. A., D. W. Smith, and D. R. Stahler. 2005. Influence of harvest, climate and wolf predation on Yellowstone elk, 1961–2004. *Oikos* 111(2): 259–270.

Wilmers, C. C., and W. M. Getz. 2005. Gray wolves as climate change buffers in Yellowstone. *PLoS Biology* 3(4): 0571–0576.

Wilmers, C. C., R. L. Crabtree, D. W. Smith, K. M. Murphy, and W. M. Getz. 2003a. Trophic facilitation by introduced top predators: Grey wolf subsidies to scavengers in Yellowstone National Park. *Journal of Animal Ecology* 72(6): 909–916.

Wilmers, C. C., D. R. Stahler, R. L. Crabtree, D. W. Smith, and W. M. Getz. 2003b. Resource dispersion and consumer dominance: Scavenging at wolf- and hunter-killed carcasses in Greater Yellowstone, USA. *Ecology Letters* 6(11): 996–1003.

Wisdom, M. J., and J. G. Cook. 2000. North American elk. In *Ecology and management of large mammals in North America*, eds. S. Demarais and P. R. Krausman, 694–735. Upper Saddle River, NJ: Prentice Hall.

Chapter 7
Carrion: It's What's for Dinner

Wolves reduce the impact of
climate change

Student Pages

Note: Reference List for Students

For more information on references cited in the Chapter 7 Student
Pages, go to teacher references on page 131.

STUDENT PAGE 7.1

Meet Dr. Chris Wilmers

Today you will learn about cutting-edge research by an ecologist named Dr. Chris Wilmers (Figure 7.5). Wilmers started studying wolves when he was a graduate student at the University of California, Davis. He was interested in interactions between wolves and their prey, and quickly realized that climate had a big impact on those interactions. In collaboration with his advisers, Drs. Wayne Getz and Eric Post, Wilmers investigated how climate change might influence the relationship between different feeding levels in the Yellowstone food web. Dr. Wilmers began working as a professor at the University of California, Santa Cruz, in January 2007, where he continues his Yellowstone research.

Figure 7.5

Dr. Wilmers uses radio telemetry to locate wolves in Yellowstone National Park.

Photograph courtesy of Chris Wilmers.

Predictions

Recall that your teacher told you wolves were eliminated from Yellowstone National Park in 1926 but were restored to the park 70 years later. When Wilmers and Getz were studying the impacts of climate change on the Yellowstone food web, they made two general predictions:

Prediction 1:   As global temperatures rise, northern and high altitude areas will experience warmer and shorter winters.

Prediction 2:   Climate change will be less harmful to ecosystems with top predators than to ecosystems without them.

Climate Data

Wilmers and Getz collected climate data from two weather stations in Yellowstone National Park (Figure 7.6).

Figure 7.6

Map of Yellowstone National Park and the Northern Range.

Most of the elk from the northern part of the park spend their winters in the Northern Range. Wilmers and Getz obtained climate data from weather stations at Mammoth Hot Springs (elevation: 1902 m) and Tower Falls (elevation: 1912 m).

Source: Modified from a map provided by the Yellowstone Wolf Project. Modified with permission.

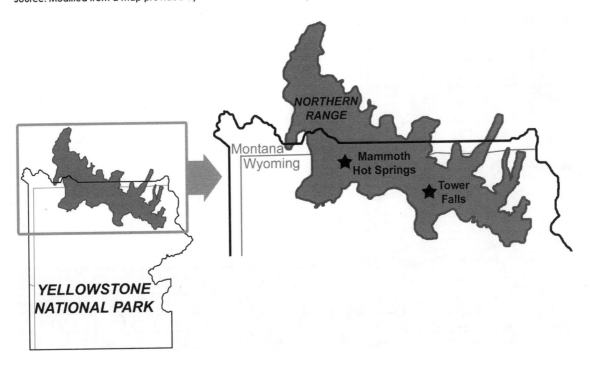

First, the scientists graphed the average monthly snow depth for each year from 1950 to 2000. Then they used a statistical technique called **regression analysis** to draw a line—called a **line of best fit**—that approximated the best overall relationship between the points. In Figure 7.7, you will see the lines of best fit for each month. Remember that the lines indicate general trends—that is, whether the average snow depth for that month has shown an overall increase, decrease, or no change during the 50-year period.

Figure 7.7

Change in snow depth from 1950 to 2000 at two weather stations in Yellowstone National Park.

Note: Each graph shows a different scale on the y-axis.

Source: Wilmers, C. C., and W. M. Getz. 2005. Modified with permission.

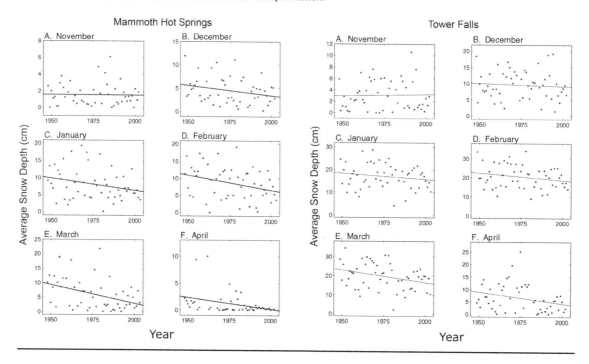

ANSWER THESE QUESTIONS:

1. How has snow depth changed from 1950 to 2000?

2. Is the trend the same or different for each weather station? Explain how.

3. a. Is the trend stronger at the beginning or the end of winter?
   b. Why might this be?

4. a. Does Figure 7.7 support Prediction 1?
   b. Why or why not?
   c. What other data would be useful? Be specific.

Wilmers and Getz also recorded the last day of snow cover at Mammoth Hot Springs and Tower Falls. For each year, they plotted the number of days it took after January 1 for the snow to melt enough to reveal bare ground. Once again, the scientists used lines of best fit to estimate the trend in snow cover for each station (Figure 7.8).

Figure 7.8

Change in length of snow season from 1950 to 2000 at two weather stations in Yellowstone National Park.

Note: Each graph shows a different scale on the y-axis.

Source: Wilmers, C. C., and W. M. Getz. 2005. Modified with permission.

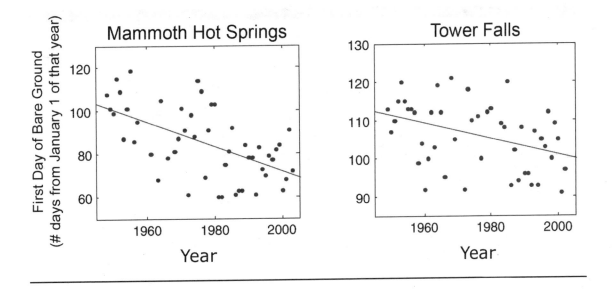

ANSWER THIS QUESTION:

1. How has the length of the snow season changed from 1950 to 2000?

Finally, the researchers graphed the number of days from January through March that temperatures were above freezing (see Figure 7.9, p. 140). TMAX is the maximum temperature in a given day.

Figure 7.9

Change in number of warm days from 1950 to 2000 at two weather stations in Yellowstone National Park.

Note: Each graph shows a different scale on the y-axis.

Source: Wilmers, C. C., and W. M. Getz. 2005. Modified with permission.

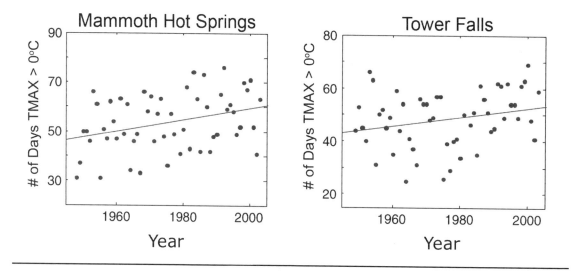

ANSWER THESE QUESTIONS:

1. How has the number of warm days changed from 1950 to 2000?

2. a. When considered together, do Figures 7.7, 7.8, and 7.9 support Prediction 1?
   b. Why or why not?
   c. What other data would be useful? Be specific.

Availability of Carrion

Winter is a tough time for large herbivores like elk. Food is in short supply and is hard to find, and it takes a lot of energy for elk just to move around in deep snow. The result is that many elk starve to death. In elk populations with no predators, deep snow is one of the main reasons elk die in the winter. However, what's bad for elk is good for ravens, eagles, coyotes, and bears. **Carrion** (rotting flesh) is a vital food source for these scavengers.

Carrion is a vital food source for these scavengers in Yellowstone National Park.

The researchers used a simple equation to estimate the amount of carrion that was available to scavengers when an elk died before wolves were reintroduced to the park in 1995. The equation describes a line of best fit constructed in the same general way as for the climate data above. In this case, the researchers started with a scatter plot showing snow depth on the x-axis and the amount of carrion on the y-axis. The data used to construct Equation 1 were drawn from a study on coyotes done before wolves were reintroduced.

Equation 1: $\rangle$ Amount of carrion without wolves $\rangle$ = (21.04 × Snow depth) + (−14.48)

Also expressed as:

$\rangle$ $C_{without\ wolves}$ $\rangle$ = (21.04 × S) + (−14.48)

(Does Equation 1 look familiar? It should, because you've used it in math class before. It's $y = mx + b$!)

Fill in the values for $C_{without\ wolves}$ in Table 7.3.

Table 7.3

Estimate of the amount of carrion available to scavengers per week with no wolves present in Yellowstone National Park.

S (Snow depth)	$C_{without\ wolves}$ (Amount of carrion)
15 cm	kg
10 cm	kg
5 cm	kg

ANSWER THIS QUESTION:

1. Explain the relationship between snow depth and the availability of carrion when there are no wolves present.

Before wolves were reintroduced to Yellowstone, the main cause of elk deaths during the winter was starvation due to deep snow. Once wolves were reintroduced to the park, however, predation became the main cause of death in elk. Since reintroduction, the availability of carrion for scavengers has depended on three things: snow depth, wolf-pack size, and how much of a carcass each wolf pack eats. Let's start by looking at Table 7.4, which shows how snow depth affects the eating behavior of wolves.

Table 7.4.

Proportion of elk carcass eaten by a typical wolf pack in Yellowstone National Park.

S (Snow depth)	E (Proportion eaten)
15 cm	0.29
10 cm	0.31
5 cm	0.32

ANSWER THESE QUESTIONS:

1. a. How do wolves change their eating behavior when there is less snow?
   b. Why do you think this might occur? (Hint: Think about how snow depth might affect the hunting ability of wolves. Remember that wolves weigh a lot less than elk do.)

Once we have used S to get E, we can complete Equation 2. (The number 0.68 refers to the edible proportion of each elk carcass.)

Equation 2: $$\text{Amount of carrion with wolves} = \text{Pack size} \times \text{Kill rate} \times (1 - \text{Proportion Eaten}) \times 0.68$$

Also expressed as:

$$C_{\text{with wolves}} = P \times K \times (1 - E) \times 0.68$$

In late winter, each wolf in Yellowstone kills an average of 240 kg/wolf/month (slightly more than two elk per month). More than 30,000 elk spend the summers in the park, and more than 47,000 elk spend the winters in the park and the nearby northern range.

Fill in the values for $C_{\text{with wolves}}$ in Table 7.5.

Table 7.5

Estimate of the amount of carrion available to scavengers per month with wolves present in Yellowstone National Park.

At the time of the Wilmer and Getz study, the average pack size was 11 wolves. In 2006, packs ranged from 4 to 19 animals (average = 10).

S (Snow Depth)	P (Pack Size)	K (Kill Rate; kg/wolf/month)	E (Proportion Eaten)	$C_{\text{with wolves}}$ (Amount of Carrion)
15 cm	11	240	0.29	kg
10 cm	11	240	0.31	kg
5 cm	11	240	0.32	kg

ANSWER THESE QUESTIONS:

1. Explain the relationship between snow depth and the availability of carrion when there are wolves present.

2. Compare and contrast the availability of carrion at different snow depths, with and without wolves.

Modeling the Effects of Climate Change

Wilmers and Getz used the equations you just worked with to create a mathematical model of the change in carrion availability between 1950 and 2000. A mathematical **model** is a tool for examining how a system or process would behave differently under different conditions. In this case, the "system or process" is the availability of carrion to scavengers and the "different conditions" are climate change and the presence or absence of wolves.

The scientists included two scenarios in their model. In Scenario 1, they assumed there were no wolves in the park (which in fact was true until 1995). They selected 100 random snow depth values for each month in the winter of 1950. Then they selected 100 random snow depth values for each month in the winter of 2000. The scientists used each of these values in Equation 1. In other words, instead of calculating $C_{\text{without wolves}}$ three times, like you did in Table 7.3, they calculated $C_{\text{without wolves}}$ 1,200 times! For each run of the scenario, Wilmers and Getz calculated the difference between the amount of carrion in 1950 and the amount of carrion in 2000. Finally, they graphed those average monthly differences (see Figure 7.10, p. 144).

ANSWER THIS QUESTION:

1. How has the availability of carrion changed between 1950 and 2000?

Figure 7.10

Estimated monthly change in carrion between 1950 and 2000, in the absence of wolves.

Note: The model was run *only* for 1950 and 2000, not any of the years in between. Each point on the line represents the difference in carrion availability between the two years.

Source: Wilmers, C. C., and W. M. Getz. 2005. Modified with permission.

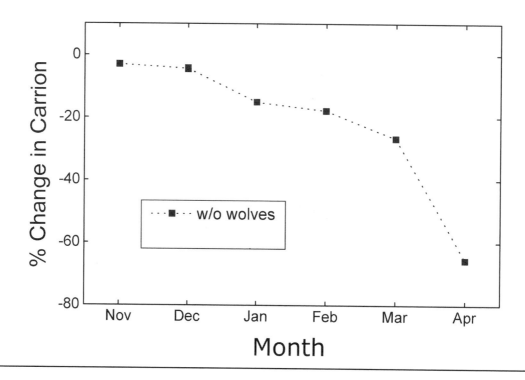

In Scenario 2, Wilmers and Getz asked, "What if?" and pretended that wolves had been present in the park for the entire 50 years. They used the same snow depth values as in Scenario 1, but also chose random wolf pack sizes from 1 to 27. Once again, they calculated $C_{with\ wolves}$ 1,200 times using Equation 2, calculated the differences between 1950 and 2000, and graphed the average monthly differences (see Figure 7.11).

Figure 7.11

Estimated monthly change in carrion from 1950 to 2000, with and without wolves.

Note: The model was run *only* for 1950 and 2000, not any of the years in between. Each point on the line represents the difference between the two years.

Source: Wilmers, C. C., and W. M. Getz. 2005. Modified with permission.

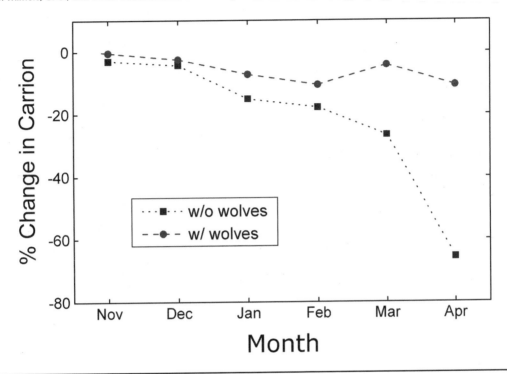

ANSWER THESE QUESTIONS:

1. If wolves had not been eradicated from Yellowstone, how would the availability of carrion have changed between 1950 and 2000?

2. a. Does this model support Prediction 2?
   b. Why or why not?
   c. What other data would be useful? Be specific.

3. If wolves had not been reintroduced to Yellowstone, how do you think climate change (and resulting shorter winters) would affect eagles, bears, and other scavengers (who depend on carrion) in the future?

Chapter 8
Right Place, Wrong Time

Phenological mismatch in the Mediterranean

Teacher Pages

AT A GLANCE

Songbirds tend to breed at the same time their primary prey is most abundant. Climate warming appears to be disrupting this match, causing reproductive failures in some species. Scientists have detected the consequences of warming for birds primarily through correlational studies. In this activity, students work in small groups and as a class to investigate "correlation versus causation." (Class time: 1–2.5 hours)

"A cardinal, whistling spring to thaw but later finding himself mistaken, can retrieve his error by resuming his winter silence. A chipmunk, emerging for a sunbath but finding a blizzard, has only to go back to bed. But a migrating goose, staking 200 miles of black night on the chance of finding a hole in the lake, has no easy chance for retreat. His arrival carries the conviction of a prophet who has burned his bridges."

—Aldo Leopold, *Sand County Almanac*, 1949

INTRODUCTION

Migration is the seasonal movement of animals from one habitat to another. Animals must time their spring migrations to match the availability of food, territories, and mates in their breeding habitat (Ramenofsky and Wingfield 2007; Figure 8.1). Precise timing is particularly important for species that breed at high latitudes and altitudes, where summers are short and spring conditions are variable (Jonzén, Hedenström, and Lundberg 2007). Climate change appears to be compromising the ability of songbirds to synchronize migratory activities with resource availability (Parmesan 2006). Some species, like the pied flycatcher (Figure 8.2), are suffering population declines because they are not breeding when their preferred prey is most abundant (Both et al. 2006).

In this directed inquiry, small groups of students work cooperatively to examine the effects of climate change on the phenology (i.e., timing of life cycle events; see Chapter 3 for more details) of pied flycatchers. By the end of the lesson, students should be able to

- recognize that a mathematical association between two variables does not necessarily imply that one of the variables is causing a change in the other; and
- explain how different trophic levels in a food chain may respond differently to climate change.

Topic: Migration of birds
Go to: *www.scilinks.org*
Code: CCPP14

Figure 8.1

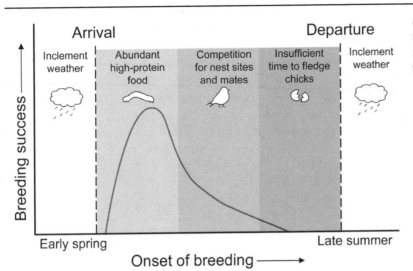

Migrating songbirds are more successful when they match their spring arrival, breeding activities, and fall departure to resource availability.

Figure 8.2

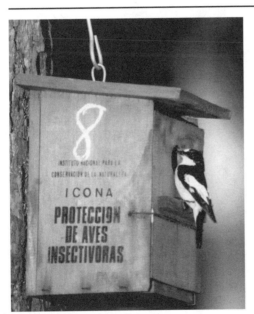

Pied flycatcher (*Ficedula hypoleuca*).

Pied flycatchers are insectivorous, cavity-nesting songbirds that breed across the Palaearctic (Europe, northern Africa, and northern Asia). They spend their winters in forests and savannahs in west Africa, south of the Sahara Desert. Adults weigh 12–13 g. Although some flycatchers live up to seven years, adult survival each year is only 40–50% (Lundberg and Alatalo 1992). The subjects of this activity are pied flycatchers that breed in the mountains of Spain. The Spanish pied flycatchers typically arrive in early May and start laying eggs 2–26 days later (Potti 1999). Females lay one clutch of 4–7 eggs per season (Moreno et al. 1995).

Although there are no pied flycatchers in North America, we have 67 other related species. You can find your local flycatcher on eBird at *http://ebird.org/ebird/eBirdReports?cmd=Start*.

At left, an adult male visits an artificial nest box.

Photograph courtesy of Juan José Sanz.

TIMING IS EVERYTHING

Pied flycatchers, like most long-distance migrants, use seasonal changes in day length and an internal biological clock to initiate migration. Once en route to their breeding grounds, flycatchers may use other environmental cues to refine their behavior (Ramenofsky and Wingfield 2007). For example, cold weather along the migration route tends to slow (or even reverse) their progress north, whereas warm weather tends to accelerate it (Both, Bijlsma, and Visser 2005).

When female pied flycatchers reach their breeding grounds, they usually mate and lay eggs within one or two weeks (Potti 1999). Under the most optimal conditions, chicks will hatch when caterpillars—which are high in protein and easy to digest—are most abundant (Martin 1987). In deciduous forests where flycatchers nest, the phenology of caterpillars is synchronized with the phenology of host trees (Buse et al. 1999) (Figure 8.3). The leaves of trees such as oaks become less palatable and nutritious as they mature, so there is strong selective pressure for caterpillars to hatch at the same time tree buds break open (van Asch and Visser 2007). It appears that over the last few decades, forest caterpillars have been emerging earlier and developing faster in response to warming temperatures (Both et al. 2006).

Although some populations of pied flycatchers (e.g., in Denmark) have adjusted to changes in caterpillar phenology by advancing their spring activities, others are suffering a "phenological mismatch" with their prey (Both et al. 2004). For instance, pied flycatchers in

Figure 8.3

Synchronization of oak, caterpillar, and flycatcher phenology.

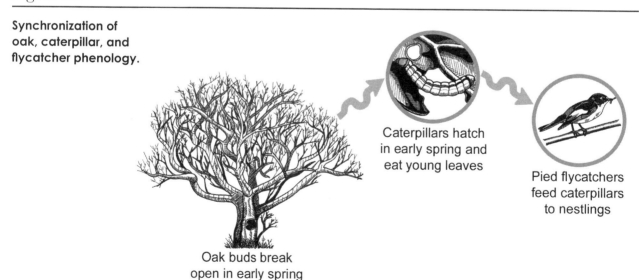

Caterpillars hatch in early spring and eat young leaves

Pied flycatchers feed caterpillars to nestlings

Oak buds break open in early spring

central Spain are breeding too late to take advantage of peak cater-pillar abundance. When clutches are laid too late, parents are forced to feed chicks prey such as grasshoppers and beetles, which are lower in nutrients and likely to slow development of chicks (Wright et al. 1998). Over the last two decades, the Spanish populations of pied flycatchers have been producing fewer and lighter **fledglings** (young birds that can fly but are still under the care of their parents).

Furthermore, the percent of fledged nestlings that return to breed in subsequent years has declined (Sanz et al. 2003). Pied flycatcher populations that suffer long-term declines in reproductive success and **recruitment** (i.e., addition of new potential breeders to the popula-tion) are likely to collapse. In The Netherlands, scientists found that flycatcher populations declined by approximately 90% over 20 years in areas where caterpillar populations peaked early relative to bird breeding activity (Both et al. 2006).

Why are some flycatcher populations able to shift migration and breeding dates when others aren't? There are several mechanisms that could account for this difference, although their relative importance is poorly understood (Pulido 2007). We list just three here. First, flycatcher populations are experiencing different regional patterns of climate change. In areas where spring warming occurs late in the season, caterpillar development is accelerated *after* birds are physi-ologically committed to breeding. Caterpillars are therefore available for a shorter period of time once chicks have hatched (Sanz et al. 2003). Second, some flycatcher populations breed in areas that experience cold snaps in early spring. When birds breed in cold weather, they have higher metabolic costs and thus less energy to invest in reproduc-tion (Stevenson and Bryant 2000). They also are at increased risk of mortality from extreme spring weather. Therefore, there may be selec-tive pressure against migrating too early (Pulido 2007). Third, migra-tory birds in poor condition are likely to arrive on breeding grounds late and start breeding late. Both climatic and nonclimatic factors that affect the body condition of birds—such as habitat fragmentation—may be overriding evolutionary changes in the timing of migration and breeding (Pulido 2007).

Activity Overview

In this directed inquiry, students use a modified Send-A-Problem framework (Barkley, Cross, and Major 2005) to examine support for the hypothesis that increasing spring temperatures are detrimental to pied flycatchers. Data sets from a study by Sanz et al. (2003) are circulated among small groups of students, who describe and inter-pret each data set in the context of the central hypothesis. Because the researchers used multiple lines of correlative evidence rather than an experimental approach, students have the opportunity to examine correlation versus causation. Please note that for simplicity we refer to the analysis used by Sanz et al. as **correlation analysis**; in fact, the researchers used regression analysis, a closely related statistical technique.

TEACHING NOTES

Materials

- Overhead transparencies of Figures 8.2–8.5 (one set per class)
- Reporting Form (Student Page 8.1; four per group)
- Consensus Form (Student Page 8.2; one per group)
- Data sets 1–6, printed on paper and on overhead transparencies (Student Page 8.3; one set per class)
- 8.5" × 11" envelopes (one each per data set) and paper clips
- Overhead projector and pens
- Blackboard, flipchart, or SMART Board
- Optional: Provide each group with a different colored piece of chalk or pen.

Procedure: Preparing for the Activity

1. Write the following hypothesis at the top of a blackboard, flipchart, or SMART Board: *High spring temperatures reduce the breeding success of pied flycatchers.* (Optional: You may wish to include the null hypothesis as well: *High spring temperatures have no effect on the breeding success of pied flycatchers.*)
2. The left side of the board will be a concept map. Write "Breeding Success" in the center of the map area. Each data set represents a concept that will be added to the map.
3. On the right side of the board, draw a table with four headings: "Data Set," "Supports Hypothesis," "Rejects Hypothesis," and "Neither Supports nor Rejects Hypothesis."

Procedure: Introducing Correlation

1. Give a brief lecture on the difference between correlation and causation:
   a. Define correlation analysis as a measure of the strength of a relationship between two variables. A positive correlation

occurs if both variables increase or decrease at the same time. A negative correlation occurs if one variable increases while the other decreases. When the results of a correlation are presented graphically, the raw data are shown as dots on a scatterplot and the relationship between the two variables as a straight line superimposed on the dots. A correlation is strong when most of the dots are tightly clustered around the line.

  b. Tell the class that correlation analysis can be used in an exploratory fashion prior to controlled experimentation or as a substitute for experiments that would be unethical or logistically difficult.

  c. Explain that without investigating the mechanism underlying a correlation, there is no way to say that the change in one variable is *causing* the change in another.

2. Illustrate the last point by having pairs of students discuss the correlations shown in Figures 8.4 and 8.5 (p. 154). (Note: Both correlations use real data!) Ask the students to consider these questions individually, then to discuss their ideas with a classmate before sharing them with the class:

  a. Is the proposed causal relationship between the x variable and the y variable biologically plausible? Why or why not?

  b. If the relationship is plausible, is it likely that x is the most important variable to influence y? What are some other important variables that influence y?

  c. If the relationship is not plausible, what are the important variables causing change in the y variable?

  d. Come up with *two* alternative explanations for the correlation between x and y.

Correlation #1: In Europe, the number of breeding pairs of storks is correlated with the human birth rate (Figure 8.4).

Interpretation: Storks deliver human babies.

Discussion points: The interpretation is not biologically plausible because, clearly, human babies do not come from storks. The activities of humans, however, may be responsible for higher stork numbers; storks appear to prefer feeding in mown grasslands (Latus and Kujawa 2005) and at open trash dumps (Tortosa, Caballero, and Reyes-López 2002).

Correlation #2: In the United States, the caloric content of dessert recipes printed in newspapers is correlated with obesity rates in major metropolitan areas (Figure 8.5).

Interpretation: The high-calorie recipes printed in newspapers are making people fat.

Figure 8.4

Human birth rates in 17 European countries.

Source: Matthews, J. 2000. How the number of births varies with stork populations in 17 European countries. *Teaching Statistics* 22(2): 36–38. Reprinted with permission of Blackwell Publishing.

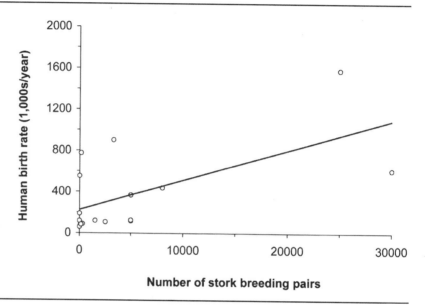

Figure 8.5

Obesity rates in 12 American cities vs. caloric content of desserts in newspapers.

Source: McCarty, C. A., D. J. McCarty, and A. C. Wetter. 2007. Newspaper dessert recipes are associated with community obesity rates. *Wisconsin Medical Journal* 106(2): 68–70. Reprinted with permission.

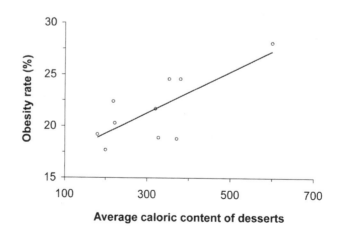

Discussion points: The interpretation is biologically plausible because excess consumption of high-calorie foods is one of the main causes of obesity. As the authors of the study point out, however, there are insufficient data to assess causality. Obese people may have a preference for high-calorie recipes and may be requesting them from their local newspaper or finding high-calorie recipes in other sources. In addition, even if recipes are one cause of obesity, they are certainly not the only cause: other diet factors, lack of exercise, and genetics are probably all more important.

Table 8.1

Internet resources.
- Photographs of pied flycatchers:
  www.stevenround-birdphotography.com/Pied%20Flycatcher.htm
- Photographs of pied flycatcher habitat in Spain:
  www.socmicolmadrid.org/imag/paiquepyrf010.jpg
- Audio of pied flycatcher song:
  www.garden-birds.co.uk/birds/piedflycatcher.htm#Voice
- Information about North American species of flycatchers:
  http://ebird.org/ebird/eBirdReports?cmd=Start

Procedure: Rotation and Synthesis

1. Introduce pied flycatchers to the class (Figure 8.2 and Table 8.1). Be sure to discuss the food chain shown in Figure 8.3.

2. Explain that researchers in Spain studied the effects of climate change on flycatchers over a period of 18 years. The study was observational, not experimental, so the researchers collected several kinds of data in order to consider multiple correlations.

3. Show Figure 8.6 to the class.
   a. Ask students to describe the two correlations and to connect both time (year) and temperature to breeding success on the concept map.
   b. Ask if, together, the two correlations support the hypothesis. Put a checkmark in the appropriate column on the board (should be in the "Supports Hypothesis" column).

4. Reinforce the idea that although increases in temperature and decreases in breeding success are occurring at the same time,

Figure 8.6

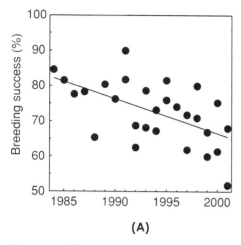

(A)

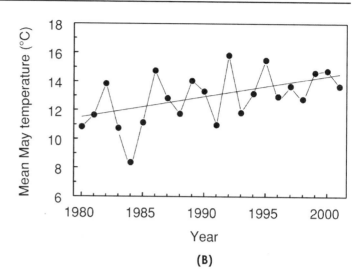

Year

(B)

Changes in (A) breeding success and (B) spring temperature over time in oak forests of central Spain.

Source: Adapted from Sanz, J. J. et al. 2003. *Global change biology.* Oxford, UK: Blackwell Publishing. Modified with permission of Blackwell Publishing.

we cannot be sure that temperature is causing the change in breeding success. Ask, "What other variables could be causing changes in breeding success over time?" Examples could include more predators, more pollution, or fewer suitable nesting trees.

5. Arrange the students into groups of two to five students. The groups should not interact with one another. If desired, assign (or allow the students to choose) some or all of the following roles to students in their groups:

    a. *Note taker*—Records the group's thoughts on each data set.
    b. *Timekeeper*—Keeps track of the time and keeps group on task.
    c. *Student representative*—Communicates with the teacher if any members of the group has questions.
    d. *Presenter*—Communicates the group's responses to the rest of the class. (Note: Although the entire group should participate in preparing the presentation, the presenter will take the lead.)
    e. *Discussion leader* (optional)—Leads class discussion at the end of the activity.

6. Give each group an envelope with one data set clipped to the outside. Groups should use a Reporting Form (Student Page 8.1) to describe the trend in the data set and give possible explanations for why that trend has occurred.

7. After 5 to 10 minutes, the group should put its completed Reporting Form in the envelope, clip the data set back to the outside of the envelope, and pass the package to a new group. In turn, they will receive a new package from another group.

8. Repeat the process three more times. Groups should use a new Reporting Form on each pass and should not look at the completed forms inside the envelopes! Note that each group will only examine four of the six data sets.

9. On the last pass, each group should examine the contents of the envelope it received, consolidate the interpretations, and complete a Consensus Form (Student Page 8.2). Students should add details or make corrections where necessary. The last group may have to choose between two or more conflicting opinions, so the presenter should be prepared to defend the group's choice! You may wish to allow additional time for this step.

Procedure: Presentations and Class Discussion

1. Each group should present its final data set to the class. Using the Consensus Form as a guide, students should point out important features of the data set and discuss the consolidated/corrected interpretations. The rest of the class should be encouraged to ask questions and challenge interpretations.

2. After a group has presented its interpretation, it should revise the concept map and put an appropriate tick mark in the table on the board. Students may change the position of previous ticks made by previous groups as the concept map is fleshed out, if the class agrees with their logic. (Optional: If you wish to keep track of the tick marks made by each group, have each group use different colored writing implements.)

3. Repeat the process until all groups have made their presentations.

4. Organize a class discussion about the concept map and table. The discussion can be led by you or the students. In the latter

case, the presenters (or discussion leaders) of each group should take charge. Give the leaders a list of questions for the class to work through, such as the following:

- Overall, did the balance of evidence support or refute the hypothesis? Explain your answer. What other data, if any, might be interesting to collect?
- Are high spring temperatures having an equal effect on all levels of this simple food chain? Why or why not? What, if any, long-term consequences might there be if spring temperatures continue to increase in this ecosystem?
- Data set 3 (peak caterpillar abundance) shows data collected in The Netherlands instead of central Spain (approximately 1,700 km away) by different investigators. Does this difference change your conclusions at all? Explain your answer.

Assessment

Because this activity relies on group work, we recommend you assess students based on their participation and interactions with their peers (Figure 8.7). To keep students on task, you may consider showing the class the assessment rubric before the activity starts. To assess students individually, you also may wish to have each student answer one of the discussion questions in paragraph form.

Figure 8.7

Assessment rubric.

Names of students in group:		
Criteria	**Points Possible**	**Points Received**
Group members worked together well.	5	
Group completed reporting sheets thoughtfully and thoroughly.	5	
Group completed consensus sheets thoughtfully and thoroughly.	5	
Group presented to the class in a way that was informative and interesting.	5	
Group contributed well to final discussion.	5	
TOTAL	**25**	
Comments:		

Modifications

If your students are familiar with correlation analysis, you could shorten this activity by skipping the "Introducing Correlation" section. You also may wish to have students write a short, individual essay summing up the important concepts of the lesson, rather than participating in a class discussion.

Europeans—especially those in northern countries—have been systematically recording spring events for centuries (Parmesan 2006). Similar studies in North America are less common, so a number of citizen science groups are encouraging the public to fill the data gap. As an extension, consider having your students contribute observations to one of the following phenology databases:

- Journey North: *www.learner.org/jnorth*
- Project BudBurst: *www.windows.ucar.edu/citizen_science/budburst*
- USA National Phenology Network: *www.uwm.edu/Dept/Geography/npn/index.html*
- Nature Watch Canada: *www.naturewatch.ca/english*

CONCLUSION

This activity could be used in an ecology or evolution unit to highlight animal behavior and the interdependence of species. However, it also is appropriate for a general science or statistics class, because it deals with correlative studies, one of the most commonly used—and misused—forms of scientific research.

REFERENCES

Barkley, E. F., K. P. Cross, and C. H. Major. 2005. *Collaborative learning techniques: A handbook for college faculty*. San Francisco: Jossey-Bass.

Both, C., A. V. Artemyev, B. Blaauw, R. J. Cowie, A. J. Dekhuijzen, T. Eeva, A. Enemar, L. Gustafsson, E. V. Ivankina, A. Järvinen, N. B. Metcalfe, N. E. I. Nyholm, J. Potti, P. A. Ravussin, J. J. Sanz, B. Silverin, F. M. Slater, L. V. Sokolov, J. Török, W. Winkel, J. Wright, H. Zang, and M. E. Visser. 2004. Large-scale geographical variation confirms that climate change causes birds to lay earlier. *Proceedings of the Royal Society B* 271(1549): 1657–1662.

Both, C., R. G. Bijlsma, and M. E. Visser. 2005. Climatic effects on timing of spring migration and breeding in a long-distance migrant, the pied flycatcher *Ficedula hypoleuca*. *Journal of Avian Biology* 36(5): 368–373.

Both, C., S. Bouwhuis, C. M. Lessells, and M. E. Visser. 2006. Climate change and population declines in a long-distance migratory bird. *Nature* 441(7089): 81–83.

Buse, A., S. J. Dury, R. J. W. Woodburn, C. M. Perrins, and J. E. G. Good. 1999. Effects of elevated temperature on multi-species interactions: The case of pedunculate oak, winter moth and tits. *Functional Ecology* 13 (Suppl. 1): 74–82.

Jonzén, N., A. Hedenström, and P. Lundberg. 2007. Climate change and the optimal arrival of migratory birds. *Proceedings of the Royal Society B* 274(1607): 269–274.

Latus, C., and K. Kujawa. 2005. The effect of land cover and fragmentation of agricultural landscape on the density of white stork (*Ciconia ciconia*) L. in Brandenburg, Germany. *Polish Journal of Ecology* 53(4): 535–543.

Lundberg, A., and R. V. Alatalo. 1992. *The pied flycatcher*. London, UK: T. & A. D. Poyser

Martin, T. E. 1987. Food as a limit on breeding birds: A life-history perspective. *Annual Review of Ecology and Systematics* 18:453–487.

Matthews, R. 2000. Storks deliver babies ($p = 0.008$). *Teaching Statistics* 22(2): 36–38.

McCarty, C. A., D. J. McCarty, and A. C. Wetter. 2007. Calories from newspaper dessert recipes are associated with community obesity rates. *Wisconsin Medical Journal* 106(2): 68–70.

Moreno, J., R. J. Cowie, J. J. Sanz, and R. S. R. Williams. 1995. Differential response by males and females to brood manipulations in the pied flycatcher: Energy expenditure and nestling diet. *Journal of Animal Ecology* 64(6): 721–732.

Parmesan, C. 2006. Ecological and evolutionary responses to recent climate change. *Annual Review of Ecology, Evolution, and Systematics* 37: 637–669.

Potti, J. 1999. From mating to laying: Genetic and environmental variation in mating dates and prelaying periods of female pied flycatchers *Ficedula hypoleuca*. *Annales Zoologici Fennici* 36(3): 187–194.

Pulido, F. 2007. Phenotypic changes in spring arrival: Evolution, phenotypic plasticity, effects of weather and condition. *Climate Research* 35(1–2): 5–23.

Ramenofsky, M., and J. C. Wingfield. 2007. Regulation of migration. *BioScience* 57(2): 135–143.

Sanz, J. J., J. Potti, J. Moreno, S. Merino, and O. Frías. 2003. Climate change and fitness components of a migratory bird breeding in the Mediterranean region. *Global Change Biology* 9(3): 461–472.

Stevenson, I. R., and D. M. Bryant. 2000. Climate change and constraints on breeding. *Nature* 406(6794): 366–367.

Tortosa, F. S., J. M. Caballero, and J. Reyes-López. 2002. Effect of rubbish dumps on breeding success in the white stork in southern Spain. *Waterbirds* 25(1): 39–43.

van Asch, M., and M. E. Visser. 2007. Phenology of forest caterpillars and their host trees: The importance of synchrony. *Annual Review of Entomology* 52: 37–55.

Visser, M. E., L. J. M. Holleman, and P. Gienapp. 2006. Shifts in caterpillar biomass phenology due to climate change and its impact on the breeding biology of an insectivorous bird. *Oecologia* 147(1): 164–172.

Wright, J., C. Both, P. A. Cotton, and D. Bryant. 1998. Quality vs. quantity: Energetic and nutritional trade-offs in parental provisioning strategies. *Journal of Animal Ecology* 67(4): 620–634.

Chapter 8
Right Place, Wrong Time

Phenological mismatch in the Mediterranean

Student Pages

Note: Reference List for Students

For more information on references cited in the Chapter 8 Student Pages, go to teacher references on page 159.

STUDENT PAGE 8.1

Reporting Form

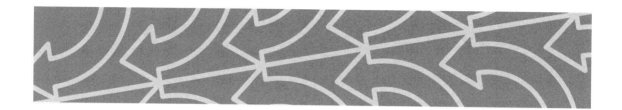

Names of people in your group:

Data set # (circle one):   1   2   3   4   5   6

Describe the relationship between the two variables in the data set.

Why do you think this relationship exists?

How is this data set related to the main hypothesis? (Circle one answer.)

Supports hypothesis   Rejects hypothesis   Neither supports nor rejects hypothesis

STUDENT PAGE 8.2

Consensus Form

Names of people in your group:

Data set # (circle one): 1   2   3   4   5   6

Examine the completed Reporting Forms in the envelope. What is the most common interpretation of the data set?

Pied flycatcher

Do you agree with this interpretation? Explain why or why not. (Hint: Think back to other data sets you looked at today.)

How is this data set related to the main hypothesis? (Circle one answer.)

Supports hypothesis   Rejects hypothesis   Neither supports nor rejects hypothesis

STUDENT PAGE 8.3

Data Sets

DATA SET 1: TIME OF LEAF DEVELOPMENT VS. YEAR IN OAK FORESTS OF CENTRAL SPAIN

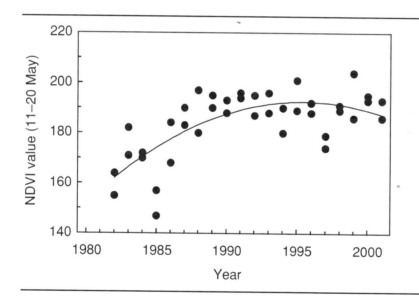

Source: Modified from Sanz, J. J. et al. 2003. *Global change biology.* Oxford, UK: Blackwell Publishing. Modified with permission of Blackwell Publishing.

Helpful hints:

- NDVI (Normalized Difference Vegetation Index) is a measurement of how much photosynthesis is occurring at the Earth's surface. In the spring, it can be used to estimate the *amount of canopy* (forest covering of leaves) that has developed and *when the canopy* developed. If the NDVI is higher in the current year than it was at the same time the previous year, it means leaves emerged earlier in the current year.

- Caterpillars hatch at the same time leaves emerge.

- Some birds may be able to use the amount of canopy as a cue to lay eggs. However, since 1980, flycatchers in central Spain *have been breeding at roughly the same time every year.*

DATA SET 2: MASS OF PIED FLYCATCHER FLEDGLINGS VS. YEAR IN OAK FORESTS OF CENTRAL SPAIN

DATA SET 3: DATE OF PEAK CATERPILLAR ABUNDANCE VS. TEMPERATURE IN OAK FORESTS OF THE NETHERLANDS

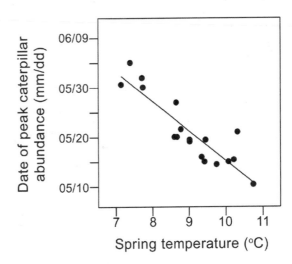

Source: Modified from Sanz, J. J. et al. 2003. *Global change biology.* Oxford, UK: Blackwell Publishing. Modified with permission of Blackwell Publishing.

Source: Modified from Visser, M. E. 2006. Shifts in caterpillar biomass phenology due to climate change and its impact on the breeding biology of an insectivorous bird. *Oecologia* 147(1): 164–172. Modified with permission of Blackwell Publishing.

Helpful hints:

- A **fledgling** is a young bird that can fly but is still under the care of its parents.
- Lighter fledglings are more likely to die than heavier fledglings.
- Light fledglings also may have a hard time migrating south to Africa, where they spend the winter.

Helpful hints:

- Caterpillars are leaf-eating machines. From the time they hatch from eggs, their main job is to accumulate enough energy to metamorphose into winged adults.
- Caterpillars are ectothermic ("cold-blooded"), like reptiles. The warmer their environment is, the faster they can move around, find food, eat, and grow.
- Some birds may be able to use insect abundance as a cue to lay eggs. However, since 1980, pied flycatchers in central Spain *have been breeding at roughly the same time every year.*

DATA SET 4: MASS OF ADULT PIED FLYCATCHERS VS. YEAR IN OAK FORESTS OF CENTRAL SPAIN

Note: Adults were weighed when their chicks were 12–13 days old.

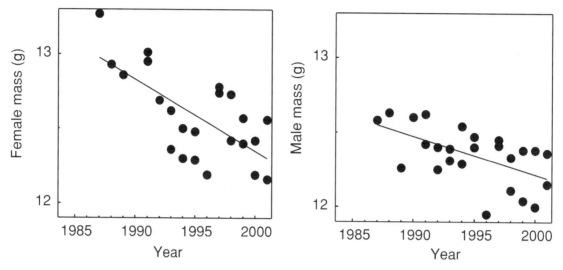

Source: Modified from Sanz, J. J. et al. 2003. *Global change biology.* Oxford, UK: Blackwell Publishing. Modified with permission of Blackwell Publishing.

Helpful hints:
- Adult birds use up a lot of energy when they are taking care of their chicks. If the adults are not getting enough to eat, they will lose weight.
- Males never sit on the eggs, but they help bring food to the chicks when they have hatched.

DATA SET 5: RECRUITMENT RATE OF PIED FLYCATCHERS VS. YEAR IN OAK FORESTS OF CENTRAL SPAIN

DATA SET 6: SPRING PRECIPITATION VS. YEAR IN OAK FORESTS OF CENTRAL SPAIN

Note: The different colored dots are from two different populations of flycatchers.

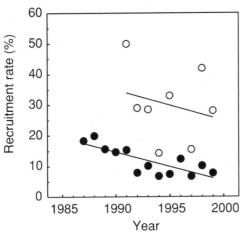

Source: Modified from Sanz, J. J. et al. 2003. *Global change biology*. Oxford, UK: Blackwell Publishing. Modified with permission of Blackwell Publishing.

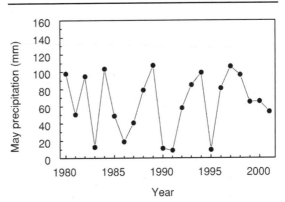

Source: Modified from Sanz, J. J. et al. 2003. *Global change biology*. Oxford, UK: Blackwell Publishing. Modified with permission of Blackwell Publishing.

Helpful hints:
- Small-bodied birds are more likely to die if spring and summer weather is unusually dry and hot or unusually wet and cold.
- Adult birds may not give their chicks enough care in dry and hot conditions.
- Caterpillars may not get enough to eat during extremely dry weather because oak leaves become distasteful and hard to chew. (Imagine someone swapping out your steak for shoe leather!)

Helpful hints:
- A **fledgling** is a young bird that can fly but is still under the care of its parents.
- The **recruitment rate** is the percent of fledglings that survive the migration to and from their wintering grounds and return to breed. (Flycatchers usually breed where they were hatched.)
- Fledglings that are in good body condition before they start their first fall migration are more likely to return the following year.

Chapter 9
Ah-Choo!

Pollen allergies increase in the Northern Hemisphere

Teacher Pages

AT A GLANCE

Pollen allergies are becoming more prevalent globally, in part because of the effect of climate change on pollen-producing plants. In this problem-based learning activity, students assume the role of public relations specialists contracted to communicate the link between climate change and allergies. This activity focuses on the importance of scientific skills to careers outside science. (Class time: 2–3.5 hours minimum)

This chapter is modified from Constible, J., L. Sandro, and R. Lee. 2008. *Journal of College Science Teaching* 37 (4): 82–89.

"There is a distinct difference between summarizing a collection of facts and telling an exciting and interesting story."

—Robert R. H. Anholt, "Dazzle 'Em With Style: The Art of Oral Scientific Presentation," 1994

INTRODUCTION

Ah, spring. Blue skies, singing birds, and ah…ah…CHOO! Pollen! Pollen is a major risk factor for respiratory allergies and asthma, which together affect 10% to 50% of people worldwide (ARIA n.d.; Bousquet, Dahl, and Khaltaev 2007). The true prevalence of chronic respiratory disease is difficult to measure: Allergies and asthma are underreported, and reporting methods are inconsistent across health agencies. Several national and international agencies, however, report that allergies and asthma are becoming more prevalent (Bousquet, Dahl, and Khaltaev 2007). Although declines in indoor and outdoor air quality are largely to blame, a growing body of evidence suggests the upswing also is related to climate change. In today's warmer, more CO_2-rich atmosphere, allergenic plants seem to be producing increasing quantities of more highly allergenic pollen, over more of the year (Beggs and Bambrick 2005).

In this problem-based learning activity, students assume the roles of public relations experts working for a firm that specializes in scientific issues. Teams of students design communication products illustrating links between climate change and pollen allergies. You can use this realistic activity to illustrate or reinforce that science process skills are relevant to an array of professions. By the end of the activity, students should be able to

- demonstrate their ability to work effectively as a group,
- interpret and prioritize tabular data of varying quality and public health significance,
- tell a story through graphics and text appropriate for use with a lay audience, and
- review the work of their peers in a thorough and unbiased manner.

GESUNDHEIT!

When pollen of an allergenic plant contacts mucus membranes in the nose, mouth, or lungs, it releases water-soluble **allergens**. In an allergic person, the allergens trigger an inflammatory response called **allergic rhinitis** ("hay fever") that can include sneezing and itching. Allergic rhinitis is serious business in the United States, costing the health care system more than $4.5 billion and resulting in 3.8 million missed work and school days per year. Allergic rhinitis also can be associated with other diseases, most notably asthma (AAAAI 2000). In this activity, students work with data on four allergenic groups of plants (Table 9.1).

Table 9.1

Allergenic plants used in "Ah-Choo!" activity.

Taxon	Clinical Importance
Birch (*Betula* spp.)	Main cause of seasonal allergies in Europe. Important allergen in Canada and the northeastern U.S.
Oak (*Quercus* spp.)	Significant allergen throughout the U.S., in southeastern Canada, and some parts of Europe.
Ragweed (*Ambrosia* spp.)	Most important pollen allergen in North America. Only recently a public health risk in Europe.
Grasses (Poaceae)	Primary pollen allergen in western Europe. Second only to ragweed in North America.

Sources: Corden and Millington 1999; Emberlin et al. 2002; Lewis, Vinay, and Zenger 1983; Taramarcaz et al. 2005.

FASTER, LONGER, AND MORE SEVERE

Increases in air temperature and atmospheric carbon dioxide (CO_2) can have positive effects on food crops such as rice. Temperature is the primary determinant of the metabolic rate of plants and therefore is an important determinant of phenology—the timing of life cycle events such as flowering (Linderholm 2006). Carbon dioxide stimulates photosynthesis and increases water use efficiency in many plants when nutrients such as nitrogen are plentiful (Bazzaz 1990). However, temperature and CO_2 act indiscriminately, promoting greater productivity in both crops and allergenic plants (Ziska et al. 2003).

The pollen season has the potential to become more severe for three reasons. First, the timing and duration of the season will shift. In the last few decades, warming has resulted in an earlier spring and a later fall across most of the Northern Hemisphere (Schwartz, Ahas, and Aasa 2006). In Europe, the pollen season has advanced by up to three weeks for some allergenic species (Student Page 9.5, Data Table 1). The pollen season also is getting longer, especially for species blooming later in the year (Student Page 9.5, Data Table 2). Second, pollen production will increase. Long-term monitoring in Europe suggests that pollen counts have increased significantly since 1969 (Student Page 9.5, Data Table 3). Likewise, experiments with two species of ragweed suggest that pollen production increases under high temperature/high CO_2 conditions as a result of more stems, more flowers, and more pollen grains per flower (Student Page 9.5, Data Table 4). Pollen allergies are usually dose-dependent: the more pollen, the greater the allergic response. Finally, pollen likely will become more allergenic. The amount of allergen in pollen varies geographically, temporally, and genetically, but increases in CO_2 and temperature appear to increase the amount of allergen per pollen grain (Student Page 9.5, Data Table 5). In related research, CO_2-enriched poison ivy plants grew twice as large as control plants and contained a more allergenic form of the compound that causes skin rashes (Mohan et al. 2006).

TEACHING NOTES

Prior Knowledge

Before the activity starts, students should understand that our climate is warming as a result of an increase in CO_2 and other greenhouse gases. You may wish to show the clip from the documentary *An Inconvenient Truth*, in which Al Gore discusses the relationship between CO_2 and global temperatures. As an alternative, you could show side-by-side graphs of surface temperature and the concentration of CO_2 and ask students to describe the relationship between the two lines (e.g., *http://earthobservatory.nasa.gov/Library/GlobalWarmingUpdate/ Images/CO2_temperature_rt.gif; http://rst.gsfc.nasa.gov/Sect16/Web-20vosto4k.jpg*).

Materials

- Scenario (Student Page 9.1; one copy per group or one overhead per class)
- Procedure (Student Page 9.2; one copy per group)
- Planning Worksheet (Student Page 9.3; one copy per group)
- Peer Review Form (Student Page 9.4; one copy per student

plus one additional copy per group for instructor)
- Data Folder and Internet Resources (Student Page 9.5; one copy per group)
- Access to computers, printers, and/or art materials

Procedure: Meeting Schedule

As a guideline for students, we split this activity into six "meetings" (Table 9.2). At a minimum, Meetings 1, 4, and 6 should be completed in class. Use of additional class time is up to you.

Table 9.2

Activity overview.

SEQUENCE	LOCATION	ACTIVITIES
Introduction and Meeting 1	In class (50–90 minutes)	Instructor facilitates class discussion about CO_2 and global temperature. Students split into small groups (two to four students) and receive scenario and data folder. Teams evaluate data, choose target audience, and brainstorm potential main messages.
Meeting 2	Outside class	Teams choose main message and type of communication product. Teams assess information needs.
Meeting 3	Outside class	Teams outline and produce drafts of products.
Meeting 4	In class (20–30 minutes)	Each team reviews one other team's product.
Meeting 5	Outside class	Teams revise products according to reviewer suggestions.
Meeting 6	In class (50–90 minutes)	Teams present products to entire class. Peer review procedure is repeated. Instructor evaluates products.

Assessment

There are multiple ways to prioritize the data, but students should be prepared to defend their rankings. When students make their final presentations, you should look for reasoning about the strength of the scientific evidence and relevance to each group's target audience. For example, Data Table 6 (Student Page 9.5) suggests that birch trees are producing more allergenic pollen in warmer microhabitats but does not rule out the effect of other environmental factors (e.g., soil type, slope, and aspect). Although the data might be relevant to some audiences, the study results are not that convincing; therefore, the table should not be top ranked.

You and the students can use the same Peer Review Form (Student Page 9.4) to assess the group projects. You should fill in the "Points" column of the Peer Review Form before duplicating the form for the class. We recommend that students receive a grade for this project based on a combination of a group score and an individual score. The individual score can be drawn from the quality of an individual's Peer Review Form, self- or peer evaluations of the group process, or your evaluations of research notes collected by each student (Barkley, Cross and Major 2005).

CONCLUSION

This open-ended activity uses a realistic scenario, current scientific data, and a structured problem-solving process to illuminate the importance of scientific skills to nonscience careers. Furthermore, most students have experienced allergic rhinitis or know someone similarly afflicted. By investigating the link between climate change and allergies, students may recognize that climate change is a present-day problem, rather than a future concern.

REFERENCES

Ahlholm, J. U., M. L. Helander, and J. Savolainen. 1998. Genetic and environmental factors affecting the allergenicity of birch (*Betula pubescens* ssp. *czerepanovii* [Orl.] Hämet-Ahti) pollen. *Clinical and Experimental Allergy* 28(11): 1384–1388.

Allergic Rhinitis and Its Impact on Asthma (ARIA). n.d. ARIA teaching slides (PowerPoint download). Retrieved September 27, 2007, from *www.whiar.org/docs.html*

American Academy of Allergy Asthma & Immunology (AAAAI). 2000. *The allergy report*, Volumes 1 and 2. Retrieved September 27, 2007, from *www.aaaai.org/ar*

Barkley, E. F., K. P. Cross, and C. H. Major. 2005. *Collaborative learning techniques: A handbook for college faculty*. San Francisco: Jossey-Bass.

Bazzaz, F. A. 1990. The response of natural ecosystems to the rising global CO_2 levels. *Annual Review of Ecology and Systematics* 21: 167–196.

Beggs, P. J., and H. J. Bambrick. 2005. Is the global rise of asthma an early impact of anthropogenic climate change? *Environmental Health Perspectives* 113(8): 915–919.

Bortenschlager, S., and I. Bortenschlager. 2005. Altering airborne pollen concentrations due to the Global Warming. A comparative analysis of airborne pollen records from Innsbruck and Obergurgl

(Austria) for the period 1980–2001. *Grana* 44(3): 172–180.

Bousquet, J., R. Dahl, and N. Khaltaev. 2007. Global alliance against chronic respiratory diseases. *Allergy* 62(3): 216–223.

Clot, B. 2003. Trends in airborne pollen: An overview of 21 years of data in Neuchâtel (Switzerland). *Aerobiologia* 19(3–4): 227–234.

Corden, J., and W. Millington. 1999. A study of *Quercus* pollen in the Derby area, UK. *Aerobiologia* 15(1): 29–37.

Emberlin J., M. Detandt, R. Gehrig, S. Jaeger, N. Nolard, A. Rantio-Lehtimäki. 2002. Responses in the start of *Betula* (birch) pollen seasons to recent changes in spring temperatures across Europe. *International Journal of Biometeorology* 46(4): 159–170.

Frei, T., and R. M. Leuschner. 2000. A change from grass pollen induced allergy to tree pollen induced allergy: 30 years of pollen observation from Switzerland. *Aerobiologia* 16(3–4): 407–416.

Frenguelli, G. 2002. Interactions between climatic changes and allergenic plants. *Monaldi Archives of Chest Disease* 57(2): 141–143.

Lewis, W. H., P. Vinay, and V. E. Zenger. 1983. *Airborne and allergenic pollen of North America.* Baltimore: Johns Hopkins University Press.

Linderholm, H. W. 2006. Growing season changes in the last century. *Agricultural and Forest Meteorology* 137(1–2): 1–14.

Mohan, J. E., L. H. Ziska, W. H. Schlesinger, R. B. Thomas, R. C. Sicher, K. George, and J. S. Clark. 2006. Biomass and toxicity responses of poison ivy (*Toxicodendron radicans*) to elevated atmospheric CO_2. *Proceedings of the National Academy of Sciences of the United States of America* 103(24): 9086–9089.

Rasmussen, A. 2002. The effects of climate change on the birch pollen season in Denmark. *Aerobiologia* 18(3–4): 253–265.

Schwartz, M. D., R. Ahas, and A. Aasa. 2006. Onset of spring starting earlier across the Northern Hemisphere. *Global Change Biology* 12(2): 343–351.

Singer, B. D., L. H. Ziska, D. A. Frenz, D. E. Gebhard, and J. G. Straka. 2005. Increasing Amb a 1 content in common ragweed (*Ambrosia artemisiifolia*) pollen as a function of rising atmospheric CO_2 concentration. *Functional Plant Biology* 32(7): 667–670.

Spieksma, F. T. M., J. M. Corden, M. Detandt, W. M. Millington, H. Nikkels, N. Nolard, C. H. H. Schoenmakers, R. Wachter, L. A. de Weger, R. Willems, and J. Emberlin. 2003. Quantitative trends in annual totals of five common airborne pollen types (*Betula, Quercus, Poaceae, Urtica,* and *Artemisia*) at five pollen-monitoring stations in western Europe. *Aerobiologia* 19(3–4): 171–184.

Taramarcaz, P., C. Lambelet, B. Clot, C. Keimer, and C. Hauser.

2005. Ragweed (*Ambrosia*) progression and its health risks: Will Switzerland resist this invasion? *Swiss Medical Weekly* 135(37–38): 538–548.

Thibaudon, M., R. Outteryck, and C. Lachasse. 2005. Bioclimatologie et allergie. *Revue Française d'Allergologie et d'Immunologie Clinique* 45(6): 447–455.

Van Vliet, A. J. H., A. Overeem, R. S. De Groot, A. F. G. Jacobs, and F. T. M. Spieksma. 2002. The influence of temperature and climate change on the timing of pollen release in The Netherlands. *International Journal of Climatology* 22(14): 1757–1767.

Wan, S. Q., T. Yuan, S. Bowdish, L. Wallace, S. D. Russell, and Y. Q. Luo. 2002. Response of an allergenic species, *Ambrosia psilostachya* (Asteraceae), to experimental warming and clipping: Implications for public health. *American Journal of Botany* 89(11): 1843–1846.

Wayne, P., S. Foster, J. Connolly, F. Bazzaz, and P. Epstein. 2002. Production of allergenic pollen by ragweed (*Ambrosia artemisiifolia* L.) is increased in CO_2-enriched atmospheres. *Annals of Allergy, Asthma and Immunology* 88(3): 279–282.

World Health Organization (WHO). 2003. *Phenology and human health: Allergic disorders. Report on a WHO meeting, Rome, Italy.* Retrieved September 27, 2007, from *www.euro.who.int/document/e79129.pdf*

Ziska, L. H., and F. A. Caulfield. 2000. Rising CO_2 and pollen production of common ragweed (*Ambrosia artemisiifolia*), a known allergy-inducing species: Implications for public health. *Australian Journal of Plant Physiology* 27(10): 893–898.

Ziska, L. H., D. E. Gebhard, D. A. Frenz, S. Faulkner, B. D. Singer, and J. G. Straka. 2003. Cities as harbingers of climate change: Common ragweed, urbanization, and public health. *Journal of Allergy and Clinical Immunology* 111(2): 290–295.

Chapter 9
Ah-Choo!

Pollen allergies increase in the
Northern Hemisphere

Student Pages

Note: Reference List for Students

For more information on references cited in the Chapter 9 Student
Pages, go to teacher references on page 176.

STUDENT PAGE 9.1

Scenario

You work for ScienceSpeak, a public relations firm that educates the public about scientific issues. Your company has won a contract with the World Health Organization (WHO) to supply materials for their new multimedia public health campaign about climate change. The WHO is specifically interested in the relationship between climate change and the increasing prevalence of allergies and asthma worldwide.

Your boss calls a meeting to discuss the contract. She gives you data folders, prepared by two expert scientists, which contain summaries of recent evidence on the effects of increasing carbon dioxide and temperature on allergenic plants. Your job is to design and produce a communication product such as a brochure, poster, web page, or television program that informs the public about potential links between climate change and allergies.

STUDENT PAGE 9.2

Procedure

Note: The meeting descriptions are intended as guides only. Be prepared to spend time on this project outside of class and to have additional meetings, if necessary.

Meeting 1: Evaluate Data

1. Examine the data prepared by your company's experts. Write a summary statement (1–2 sentences) to describe the main trend for each data set in the Planning Worksheet (Student Page 9.3). Remember that the public is interested in the *big* picture.

2. Choose your target audience. For example, are you trying to reach children or adults? People who live in the city or in the country?

3. Use the Planning Worksheet to rank your summary statements in order of importance (1 = most important, 6 = least important). For each data table, consider:

a. *The strength of the scientific evidence.* For example: How were the data collected? Are there sufficient data to support a strong claim? How consistent is a response across plant species? Across geographical areas?

b. *The relevance of the data to your target audience.* For example: Where does your audience live? What plant species might they encounter? What sort of professions or recreational activities are your audience members engaged in?

4. Give your reasoning for the ranking of the top two statements.

5. On the back of the Planning Worksheet, brainstorm possible messages for your communication product.

Meeting 2: Generate Strategies

1. Use the results of your brainstorming session at the end of Meeting 1 to decide what your main message will be. Remember, a respected international organization will be using your product to inform the public about an important issue. Make sure your message is clear, factual, scientifically accurate, and catchy.

2. Decide how you will communicate your message to your target audience. You may choose a product from this list:
   - Brochure
   - Poster
   - Web page
   - Television program (you may provide the script or a video)
   - Another idea (Get your instructor's approval.)

3. Assess your knowledge by answering these questions:
   a. What do we know?
   b. What do we need to know?
   c. How can we find out what we need to know?

4. Before your next meeting, you will have to collect the information you need to complement the data provided by the experts. Decide what tasks need to be completed before the next meeting. Make sure everyone gets an equal amount of work.

Meeting 3: Produce Draft of Product

1. Review what each team member has learned since Meeting 2.

2. Outline your product. At a *minimum*, your product will need to include the following:
   a. An introduction that catches the reader's/viewer's interest and gets your message across.
   b. A summary of your top-ranked data that includes at least two graphics. Even within a given data set, not all the information is equally important. You will need to emphasize the critical points. You should include at least one graph, but you also may use maps, flowcharts, diagrams, or other graphics.
   c. A summary of your less important data.
   d. A conclusion that reinforces your message and ties loose ends together.

3. Produce a draft of your product, based on the outline you made in Step 2. You will have time to revise your draft later.

Meeting 4: Peer Review

Scientists (and many other professionals) use a process to improve their work called *peer review*. Before a scientific document can be published, it must be approved by a panel of fellow scientists. The procedure you are about to follow incorporates elements of a scientific peer review:

1. Your team will present its work to another team to be evaluated. Give one copy of the Peer Review Form (Student Page 9.4) to each person on the other team. Make sure your group name and intended audience are filled in on the form you give to the other team.

2. Another team will be presenting its work to you. Each member of your team will provide an individual review of the product by
   a. providing positive feedback on what worked well in the product,
   b. suggesting changes that will help the product, pointing out any errors or overlooked/misinterpreted data, and
   c. thoroughly evaluating the product according to the Peer Review Form. Reviewers should be strict but fair in their assignment of points to each category.

3. Return the Peer Review Forms your team completed to your partner team. Receive their reviews of your team's work.

Meeting 5: Revision of Product

1. Incorporate the peer review suggestions you think would improve your communication product. One member of your team should keep a record of each suggestion you actually use. Effective implementation of appropriate suggestions will mean a higher score for your product.

2. Make final adjustments to your product.

Meeting 6: Presentation of Final Product

1. Present your product to the class. When the team you evaluated in Meeting 4 presents its product, repeat the peer review process.

2. Submit the Planning Worksheet, Peer Review Forms, and your final product to your instructor.

STUDENT PAGE 9.3

Planning Worksheet

Planning Worksheet (Meeting 1)			
Group Name:			
Data Set	**Summary Statement**	**Rank (1 = Most important)**	**Reason for Ranking Top Two Statements**
Table 1			1.
Table 2			
Table 3			
Table 4			2.
Table 5			
Table 6			

STUDENT PAGE 9.4

Peer Review Form

Peer Review Form (Meetings 4 and 6)				
Reviewer Name:				
Group Name:		**Intended Audience:**		
Type of presentation (circle one): Brochure Poster Web page Television program Other:				
Criteria	**Points**	**Peer Review (Draft)**	**Peer Review (Final Product)**	**Instructor (Final Product)**
Introduction				
The main message is clear, factual, scientifically accurate, and catchy.				
Body				
The product summarizes all six data tables.				
Appropriate types of figures (e.g., graphs, maps, etc.) are used to highlight the most important data. The figures are detailed and creative.				
Conclusions				
Appropriate conclusions are made based on the available data.				

(continued)

(continued from p.185)

Criteria	Points	Peer Review (Draft)	Peer Review (Final Product)	Instructor (Final Product)
General Criteria				
The text/spoken word is easy to read/hear.				
The product is organized logically.				
The format and language are appropriate to the target audience.				
The product is an appropriate length.				
There are no errors in spelling and/or grammar.				
TOTAL				

List three specific things you liked about the draft communication product:

1.

2.

3.

List three specific suggestions for improving the draft communication product:

1.

2.

3.

General comments:

STUDENT PAGE 9.5

Data Folder and Internet Resources

Data Table 1: Observational Data.
Change in start of pollen season in Europe from 1969–2003. "--" = No data.

Country	Average change in start of pollen season			
	Oak	Birch	Ragweed	Grass
Austria	--	17 days earlier	--	--
Belgium	--	23 days earlier	--	--
Denmark	--	13 days earlier	--	--
France	12 days earlier	6 days earlier	NO CHANGE	NO CHANGE
Italy	8 days earlier	--	--	6 days earlier
The Netherlands	18 days earlier	10 days earlier	--	6 days earlier
Switzerland	21 days earlier	20 days earlier	20 days earlier	14 days earlier
United Kingdom	19 days earlier	--	--	--

Source: Data compiled from Clot 2003; Corden and Millington 1999; Emberlin et al. 2002; Frenguelli 2002; Rasmussen 2002; Thibaudon, Outteryck, and Lachasse 2005; Van Vliet et al. 2002.

Data Table 2: Observational Data.
Duration of pollen season (17 pollen types) in Europe (450 stations), 1974–2002.

Flowering season*	Average change in length of pollen season
Winter	8 days shorter
Early spring	3 days longer
Mid-spring	3 days longer
Late spring	4 days longer
Summer	2 days longer
Autumn	4 days longer

* Trees typically flower from winter to mid-spring. Grasses can flower all year in some areas, but usually peak from late spring to summer. Most allergenic herbs flower from summer to autumn.

Source: Data compiled from Jaeger 2001 (cited in World Health Organization 2003).

Data Table 3: Observational Data.

Change in amount of airborne pollen in Europe, 1969–2003. "--" = No data.

Country	Average change in amount of pollen		
	Oak	Birch	Grass
Austria	--	4.5 × more	2.4 ×more
Belgium	2.7 × more	2.3 × more	NO CHANGE
Denmark	--	4.1 × more	--
France	2.3 × more	5.3 ×more	--
Germany	NO CHANGE	NO CHANGE	NO CHANGE
Netherlands	4.7 × more	2.4 × more	NO CHANGE
Switzerland	2.6 × more	2.4 × more	1.1 × more
United Kingdom	4.8 × more	4.7 × more	NO CHANGE

Source: Data compiled from Bortenschlager and Bortenschlager 2005; Frei and Leuschner 2000; Rasmussen 2002; Spieksma et al. 2003; Thibaudon, Outteryck, and Lachasse 2005.

Data Table 4: Experimental Data.

Effect of increased CO_2 and temperature on pollen production in common ragweed. Half of the ragweed plants in each experiment were treated either with the predicted amount of CO_2 that may be in the atmosphere in 50 years or the predicted temperature that may occur in 50 years.

Location	Growing technique	Treatment	Observed change in pollen production in treated plants
Maryland	Environmental chambers	High CO_2	1.9 × more pollen than control plants
Illinois	Greenhouses	High CO_2	1.6 × more pollen than control plants
Oklahoma	Outdoor field	High temperature	1.8 × more pollen than control plants

Source: Data compiled from Wan et al. 2002; Wayne et al. 2002; Ziska and Caulfield 2000.

Data Table 5: Experimental Data.

Effect of CO_2 on concentration of protein allergens in ragweed pollen. Plants were grown in environmental chambers. ELISA is a laboratory test used to detect antibodies; the higher the ELISA, the higher the concentration of allergen.

Concentration of CO_2 (μmol/mol)	Concentration of allergen from ragweed plants (ELISA/mg pollen)
280 (preindustrial levels)	93
370 (current levels)	103
600 (projected future levels)	178

Source: Data compiled from Singer et al. 2005.

Data Table 6: Observational Data.
Effect of temperature on concentration of protein allergens in birch pollen. Plants were grown in outdoor gardens in northern Finland.

Location	Concentration of allergen from birch trees (% antibody bound to protein)
Hill, 270 m above sea level	103
River valley, 90 m above sea level (Daily average temperatures 1.0–2.5°C warmer than hill location)	130

Source: Data compiled from Ahlholm et al. 1998.

For more information:

- Allergen/Additive/Preservative Search: http://allallergy.net/allergensearch.cfm
- American Academy of Allergy Asthma & Immunology: *www.aaaai.org*
- American College of Allergy, Asthma & Immunology: *www.acaai.org*
- Asthma and Allergy Foundation of America: *http://aafa.org*
- National Institute of Allergy and Infectious Diseases: *http://www3.niaid.nih.gov*
- Pollen Library: *www.pollenlibrary.com*
- World Pollen Calendar: *www.hon.ch/Library/Theme/Allergy/Glossary/calendar.html*

Chapter 10
Cruel, Cruel Summer

Heat waves increase from pole to pole

Teacher Pages

AT A GLANCE

Scientists expect that a warmer climate will cause more severe, more frequent, and longer heat waves. Heat waves pose a significant health risk to everyone, but especially to poor, elderly, and chronically ill individuals. In this open-ended inquiry, students use raw data from public health websites to investigate the relationship between extreme heat and human mortality. The activity stresses data acquisition and analysis skills. (Class time: 3–4 hours minimum)

"Hot summer streets
And the pavements are burning
I sit around.
Trying to smile, but
The air is so heavy and dry."

—Bananarama, "Cruel Summer," 1983 (Reprinted with permission)

INTRODUCTION

Heat waves may seem mild mannered compared to other extreme weather events, but they are actually more dangerous than the average storm. **Heat waves** (i.e., several days of extremely hot and possibly humid weather) cause 25% of weather-related fatalities in the United States and kill 2–3 times more people than lightning, tornadoes, floods, or hurricanes (NWS 2007a). The global frequency of heat waves has increased since 1950 (Trenberth et al. 2007), and climate models predict they will become longer, more severe, and more frequent in the next century (Christensen et al. 2007).

In this open-ended inquiry, students pose as epidemiologists working for a local health agency. Students use raw data from online databases to study the relationship between climate and human mortality. By the end of the inquiry, students should be able to do the following:

- Manipulate data and generate graphs in Excel or similar software; and
- Demonstrate an understanding of techniques used in **epidemiology** by carrying out a research study on heat-related mortality.

TOO HOT TO HANDLE

Parents who have stayed up all night with a feverish child probably have the human body's normal temperature—37°C, or 98.6°F—committed to memory. When our bodies stray too far from that norm, thermoregulatory centers in our brains initiate a number of responses to warm us up or cool us down. Under normal conditions, heat loss occurs via sweat production, increased blood flow from the body's core to the skin, and an increase in respiration and heart rate. Under extreme conditions, heat loss is impaired and illness or death can result (Koppe et al. 2004).

Classic heat-related illnesses range from heat rash, a mild condition caused by excess sweating, to heat stroke, a potentially deadly condition caused by a failure in the body's ability to regulate heat. Heat-related deaths are underreported, however, as they are often confused with deaths caused by other illnesses (Koppe et al. 2004). For example, extreme heat increases the risk of a heart attack because of added strain on the heart. Unless body temperature is measured at the time of death, it's difficult to assess whether an apparent heart attack victim actually died of heat stroke (Kilbourne 1997).

WHO'S AT RISK?

Anyone can be affected by a heat-related illness, but high-risk groups (i.e., those that develop the greatest number of cases of heat-related illnesses) include individuals who are chronically ill, poor, or elderly. Chronically ill individuals may be dehydrated or under the influence of medications that impair their body's ability to regulate heat. People with chronic mental illnesses might fail to take protective measures against extreme heat (EPA 2006). Poor individuals may have limited access to central air conditioning, which can reduce the risk of heat-related mortality by 42% (Rogot, Sorlie, and Backlund 1992). Furthermore, low-income neighborhoods often are hotter than high-income neighborhoods because they are more densely settled and have less vegetation (Harlan et al. 2006). Elderly people are more sensitive to heat because of their decreased ability to sweat and weakened circulatory systems. The elderly also have an increased risk of chronic illness (both mental and physical) and may have limited income (Kilbourne 1997).

Vulnerability to heat also varies by geographic region, in part because humans can acclimatize to local conditions. People living in areas with consistently high temperatures tend to have an increased capacity to sweat, slightly lower metabolic rates, and higher rates of blood flow than people living in areas with infrequent heat (Koppe et al. 2004). In the United States, heat waves tend to have the greatest impact in the Northeast and Midwest (EPA 2006).

Finally, urban dwellers seem to have a higher risk of mortality during heat waves than rural dwellers (McGeehin and Mirabelli 2001). Cities can be anywhere from 1 to 10°C warmer than surrounding areas. This phenomenon, called the urban heat island effect (see Chapter 2), occurs because roads and other infrastructure modify the exchange of radiation and heat between the surface and the atmosphere. City centers, with their tall buildings, dark-colored and impervious surfaces, and relative lack of vegetation have a higher capacity for heat storage and a lower rate of evaporative cooling than

SC*L*INKS.
THE WORLD'S A CLICK AWAY
Topic: Heat/temperature
Go to: www.scilinks.org
Code: CCPP15

less-developed areas. Furthermore, the release of carbon dioxide and industrial pollutants in urban areas results in a temporary but significant increase in the local greenhouse effect. The temperature difference between urban and rural areas is greatest 2 to 3 hours after sunset because the narrow streets and tall buildings in cities block outgoing longwave radiation and reduce wind speeds (Grimmond 2007). The risk of death during a heat wave is much higher when nighttime temperatures remain elevated (Figure 10.1) (Karl and Knight 1997).

Figure 10.1

More than 500 people died in the 1995 heat wave in Chicago, Illinois, when extreme heat (>27°C) and humidity (>40% relative humidity) persisted for 48 hours.

Photograph courtesy of Juanita Constible.

HOT VERSUS COLD

Some researchers have suggested that as the climate warms, decreases in winter mortality will offset increases in heat wave mortality (Davis et al. 2004). However, far more deaths are caused by extreme heat than extreme cold. In a 30-year study of weather-related deaths in Germany, researchers found that the **mortality rate** (number of deaths per 100,000 people) during heat waves increased twice as much as the rate during cold spells (Laschewski and Jendritzky 2002). From 1988 to 2006, there were 2,929 heat-related deaths in the United States but only 515 cold-related deaths (NWS 2007a). Most deaths caused by winter weather are caused not by cold temperatures, but by the hazards associated with snow, ice, and strong winds (e.g., falling ice, loss of electric power, etc.) (NWS 2001; NWS 2007b). Although the number of extreme cold days has already declined, a warmer climate likely will lead to both increases in hot days in the summer and more severe storms in the winter. European climate models, for example, suggest that climate warming will result in heavier snows and more extreme winter winds over much of Europe (Beniston et al. 2007).

MAKING SCIENCE AND MATH RELEVANT

This inquiry involves a number of high-level skills and therefore may be best suited for an advanced class or for an individual science fair project. (See "Modifications," p. 204, for ways to simplify the activity.) However, we hope to engage and motivate students not primarily interested in science by doing the following:

- Identifying an authentic problem: The world's population is aging (ESA 2005) and becoming more urban (UNEP 2007), meaning heat-related mortality will increase even without the effects of climate change.
- Engaging the interest of students who live in urban and/or low income areas by focusing on issues particularly relevant to them.
- Demonstrating that science has societal value: This inquiry highlights epidemiology, a critical tool in the fight against threats such as epidemics, terrorism, and natural disasters.
- Challenging students to explore: The open-ended nature of this inquiry fosters independent thought and provides an authentic scientific experience.

This activity also is perfect for integrating math and science. It could be team taught by biology and math/statistics instructors or even used as a stand-alone activity in a statistics class.

TEACHING NOTES

Prior Knowledge

Students should understand the difference between climate and weather before they start the inquiry. You may introduce or reinforce this difference (as well as engage students with a bit of humor) by showing a relevant video clip of the *Daily Show*, available at *www. comedycentral.com/motherload/index.jhtml?ml_video=81842.*

Materials

- An overhead showing Figure 10.5 (one per class)
- Student Page 10.1: Scenario Letter (one per group, or one overhead per class)
- Student Page 10.2: Epidemiology 101 (one per student)
- Student Page 10.3: Research Questions and Expectations (one per group)
- Computers with Microsoft Excel or other appropriate spreadsheet software (one per group)

Procedure: Preparing the Data Sets

The open-ended nature of this inquiry can be daunting to students (and teachers) accustomed to highly structured activities. Students must master substantial background information and a number of computing skills to produce a polished product. However, the processes involved in this activity (data manipulation and analysis, problem solving, etc.) are more important than a final product. Because students might be frustrated by a perceived lack of progress, you may wish to assist them by preparing the data sets ahead of time:

1. Choose a study area near your school. To ensure you have sufficient data, we recommend selecting the nearest large city or your entire county. For example, in Appendix 10.1 (p. 217) we used mortality data from Baton Rouge, Louisiana, for Research Questions 1–3, but we had to use data from East Baton Rouge Parish (county) for Question 4.
2. Define a heat wave for your study area (Figure 10.2).

Figure 10.2

Defining heat waves.

1. Before you can examine the effects of heat waves on your study population, you will have to define an **extreme heat day** for your study area:
   a. Visit the United States Historical Climatology Network website (*http://cdiac.ornl.gov/epubs/ndp/ushcn/daily.html*).
   b. At the bottom of the page, click on "Go to daily data."
   c. Scroll down until you see a white map of the United States on a gray background. Click on the state containing your study area (e.g., Louisiana).

d. Click on the dot representing the weather station closest to your study area (e.g., 160549, Baton Rouge, Louisiana). Note: You also can find data for the entire state by clicking on the link at the top of the page.

e. Click on "Create a download file."

f. Put a check in the "Temperature Max" box and click the "submit" button. *Warning!* If you change the file name before you move to the next screen, your download will fail!

g. On the next screen, click the download link. A File Download window will appear. Click the "Save" button.

h. Double click on the icon of your downloaded file to open it in Excel.

i. Use the "Save As" function in Excel to save the file as a Microsoft Excel Workbook, instead of a CVS file. You may want to include the name of your study area and the phrase *heat waves* in the file name. Remember to *save your work often!*

j. Delete the first row (starts with the word *Source*). The column headings should be "Day" (day of the month), "JD" (Julian day), "Month" (month coded as numbers 1–12), "State_id" (the ID of your weather station; you can delete this column if you wish), "Year," and "TMAX" (maximum temperature in °F; you can convert this to °C if you wish).

k. We are only interested in summer months. Select all your data by clicking on the empty square in the top left-hand corner of the spreadsheet (i.e., to the left of the "A" column heading). Under the "Data" menu, choose "Sort." Sort the data by month, then delete records from months other than June, July, and August (months 6, 7, and 8).

l. Scroll down to the bottom of the spreadsheet to see how many data records you have. Excel can only calculate functions on 8,191 rows of data or less. If you have more, re-sort by year, then delete enough data from the earliest years in your records to bring the total number of rows to 8,191.

m. Put your cursor in the first empty cell at the bottom of your temperature column (F6994 in our Baton Rouge example). From the menu at the top of the screen, go to "Insert" and then "Function."

n. If you are working on a PC, type "percentile" (without the quotes) in the "Search for a function box" and click "Go." If you are working on a Macintosh, click "All" to see a list of functions.

o. Choose "PERCENTILE."

p. In the Array box, type in the range of your data. In our Baton Rouge example, the range would be F2:F6993. The "F" refers to the column containing temperature data, "2" refers to the first row containing data, and "6993" refers to the last row containing data.

q. In the K box, type "0.90" (without the quotes). Click "OK." The number you've calculated is the 90th percentile (95°F, in our Baton Rouge example). In other words, for 90% of the summer, maximum daily temperatures in your area are lower than the value you calculated. Any temperatures equal to or higher than the value you calculated are extreme events. Write this value down for safekeeping.

2. Mark the heat waves from 1996 onward. (Mortality data generally is unavailable before 1996). For the purposes of this study, a heat wave is three consecutive extreme heat days (we are ignoring humidity for simplicity's sake).

   a. Copy your original worksheet into a new worksheet within your spreadsheet.

   b. Delete the 90th percentile you calculated in Step 1q.

   c. Sort the data by maximum temperature (TMAX).

   d. Change the background color of rows containing extreme weather days ($\geq$ your 90th percentile) to a bright color. Under the "Format" menu, choose "Cells," then click the "Patterns" tab, and choose a color under "Shading."

   e. Sort the data by Year, Month, and Day.

   f. Look for periods of three or more consecutive extreme weather days by scrolling down the worksheet. For each period, type the words "HEAT WAVE" in the column to the right of the last day of the heat wave. In our Baton Rouge example, there were 21 heat waves from 1996 through 2005.

Note: The **heat index**, a measure of how hot it "feels" based on heat and humidity combined, is more biologically relevant than temperature alone. Extreme heat is more dangerous when humidity is also high because evaporative cooling is greatly reduced. However, you need either relative humidity or dew point for the calculation, and neither data set is widely available. If you are interested in learning more about the heat index, visit *www.crh.noaa.gov/jkl/?n=heat_index_calculator*.

3. Obtain mortality data for your study area (Figure 10.3).

Figure 10.3

Obtaining mortality data.

To get weekly mortality data for a major city:

1. Visit the Morbidity and Mortality Weekly Report (MMWR) website (*http://wonder.cdc.gov/mmwr/mmwrmort.asp*).
2. Select a year using the drop down menu on the left. Ignore the week menu.
3. Select a city. Cities are arranged alphabetically within regions of the country.
4. Click "Submit."
5. Drag your mouse over the MMWR weeks that include June, July, and August (usually weeks 22–35; see Figure 10.3a on page 201). MMWR weeks run from Sunday to Saturday.
6. Copy the selected data into an Excel spreadsheet. Insert a new row at the top of the spreadsheet for headings and three new columns for year, month, and starting day of each week. Save the file.
7. Repeat Steps 1–5 for at least four other years, copying each year of data into your original Excel file. You might wish to repeat the steps for all available years to ensure you have sufficient data.

To get yearly mortality data for a county:

1. Visit the CDC Wonder Compressed Mortality File website (*http://wonder.cdc.gov/mortSQL.html*).
2. Click the link to "Mortality for 1999–2005 with ICD 10 codes."
3. In Section 1, select "Group Results by County."
4. In Section 2, select the "States" button and then select your state from the "Browse" menu.
5. In Section 3, choose "All Ages," "All Years," "All Genders," "All Races," and "All Categories." You can go back later, if you wish, to do a detailed analysis by age or by some other category.
6. In Section 4, select "ICD-10 Codes." If you want a specific cause of death, use the "Browse" menu. To get data on violent crime, scroll down to "V01-Y89 (External causes of morbidity and mortality)" and click the "Open Fully" button. Select "X85-Y09 (Assault)."

7. Ignore Section 5. In Section 6, select "Calculate Rates per 10,000" and "No age-adjusted rates."
8. Ignore Section 7. Click the "Send" button.
9. On the next screen, click "Export Results" and "Save."
10. Open the resulting text file within Excel (you will need to have "All Files" selected within the "Files of type" drop-down menu). You will see a series of "Text Import Wizard" screens. Click "Next," "Next," and "Finish" to import the file into Excel. Save the file as a Microsoft Excel Workbook.

Note: You also can try your local department of health, community health, or public health for other data sets. Try these search terms on your local agency's website (without the quotes):
- "Health statistics" or "health data"
- "Mortality statistics" or "mortality data"
- "Vital statistics"

---

4. Obtain population data for your study area (Figure 10.4, p. 202).

Figure 10.3a

Year	\multicolumn{7}{c}{Starting Date of MMWR Weeks 22–28}						
	22	**23**	**24**	**25**	**26**	**27**	**28**
1996	May 26	Jun 2	Jun 9	Jun 16	Jun 23	Jun 30	Jul 7
1997	---	Jun 1	Jun 8	Jun 15	Jun 22	Jun 29	Jul 6
1998	May 31	Jun 7	Jun 14	Jun 21	Jun 28	Jul 5	Jul 12
1999	May 30	Jun 6	Jun 13	Jun 20	Jun 27	Jul 4	Jul 11
2000	May 28	Jun 4	Jun 11	Jun 18	Jun 25	Jul 2	Jul 9
2001	May 27	Jun 3	Jun 10	Jun 17	Jun 24	Jul 1	Jul 8
2002	May 26	Jun 2	Jun 9	Jun 16	Jun 23	Jun 30	Jul 7
2003	---	Jun 1	Jun 8	Jun 15	Jun 22	Jun 2	Jul 6
2004	May 30	Jun 6	Jun 13	Jun 20	Jun 27	Jul 4	Jul 11
2005	May 29	Jun 5	Jun 12	Jun 19	Jun 26	July 3	Jul 10
2006	May 28	Jun 4	Jun 11	Jun 18	Jun 25	Jul 2	Jul 9
2007	May 27	Jun 3	Jun 10	Jun 17	Jun 24	Jul 1	Jul 8
2008	---	Jun 1	Jun 8	Jun 15	Jun 22	Jun 29	Jul 6

Year	\multicolumn{7}{c}{Starting Date of MMWR Weeks 29–35}						
	29	**30**	**31**	**32**	**33**	**34**	**35**
1996	Jul 14	Jul 21	Jul 28	Aug 4	Aug 11	Aug 18	Aug 25
1997	Jul 13	Jul 20	Jul 27	Aug 3	Aug 10	Aug 17	Aug 24
1998	Jul 19	Jul 26	Aug 2	Aug 9	Aug 16	Aug 23	Aug 30
1999	Jul 18	Jul 25	Aug 1	Aug 8	Aug 15	Aug 22	Aug 29
2000	Jul 16	Jul 23	Jul 30	Aug 6	Aug 13	Aug 20	Aug 27
2001	Jul 15	Jul 22	Jul 29	Aug 5	Aug 12	Aug 19	Aug 26
2002	Jul 14	Jul 21	Jul 28	Aug 4	Aug 11	Aug 18	Aug 25
2003	Jul 13	Jul 20	Jul 27	Aug 3	Aug 10	Aug 17	Aug 24
2004	Jul 18	Jul 25	Aug 1	Aug 8	Aug 15	Aug 22	Aug 29
2005	Jul 17	Jul 24	Jul 31	Aug 7	Aug 14	Aug 21	Aug 28
2006	Jul 16	Jul 23	Jul 30	Aug 6	Aug 13	Aug 20	Aug 27
2007	Jul 15	Jul 22	Jul 29	Aug 5	Aug 12	Aug 19	Aug 26
2008	Jul 13	Jul 20	Jul 27	Aug 3	Aug 10	Aug 17	Aug 24

Figure 10.4

Obtaining population data.

To get population estimates for 2000 to the present:
1. Visit the present U.S. Census website *(www.census.gov/ popest/cities)*.
2. Under "Popular Tables," choose "All Incorporated Places" and click "Go."
3. Download the Excel file for your state.

For population estimates before 2000, try these options:
- State Data Resources Program: *www.census.gov/sdc/www*
- Search the internet for your place name, the year of interest, and the words *census* or *population*.
- Extrapolate earlier years based on the population trend from 2000 to the present

5. Summarize the heat wave, mortality, and population data in one spreadsheet. We recommend that, at minimum, you use the following columns:
   a. Mean heat wave TMAX (see Figure 10.2, Step 2) in each week of June, July, and August for each year (index of heat wave severity).
   b. Number of heat waves per month in each of June, July, and August for each year (index of heat wave frequency).
   c. Average duration of heat waves per month in each of June, July, and August for each year (index of heat wave duration).
   d. Mortality rate in each week of June, July, and August for each year from:
      - all causes,
      - infectious diseases (pneumonia and influenza are available by week, other diseases by year), and
      - assaults (only available by year, not by week).
   e. Population of study area in each year.

Please note these instructions are for locations within the United States. Similar resources in other countries can be found on the internet.

Procedure: Preparing the Student Pages

1. Complete the blanks on the Scenario Letter (Student Page 10.1) before copying it for the students. You can find the

202

name, website, and logo of your local public health agency at the Association of State and Territorial Health Officials website (*www.astho.org/index.php?template=regional_links.php&PHPSESSID=8a104e76f1b4d4d321337b835a8fc17b*).

2. Copy the Research Questions (Student Page 10.3) for the students (see Appendix 10.1, page 217, for sample solutions to these questions). You might want to add questions or modify the existing ones. For example, students could collect and manipulate their own data to answer questions such as "How do indoor temperatures of buildings in our area change with the weather?" A team of students could answer this question by deploying thermometers or automated temperature loggers in and around the school buildings, their houses, and/or their parents' places of employment.

Procedure: Starting the Activity

1. Show Figure 10.5 to the class. Ask your students to describe the pattern in each part of the figure. Does this figure represent changes in weather or climate? What implications,

Figure 10.5

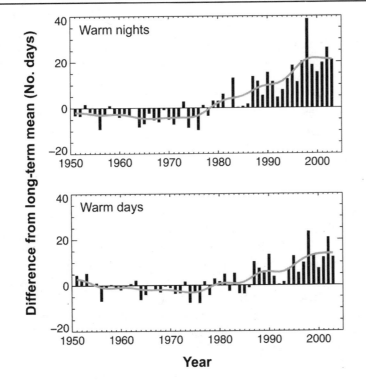

Global trend in warm days and nights, 1950–2003.

Source: Alexander, L. V., et al. 2006. © 2006, American Geophysical Union (AGU). Modified with permission from AGU.

if any, does the pattern have for local weather patterns?

2. Form small student teams. It may be helpful to define roles for each student, such as reader, writer, data manipulator, or group leader.

3. Hand out the scenario or display it on an overhead (Student Page 10.1). After students have read the scenario, explain that scientists who work on public health concerns such as the one in the letter are called epidemiologists. Give a short lecture based on Epidemiology 101 (Student Page 10.2) or have students read the information. (Optional: Invite a public health professional to give a guest lecture on epidemiology to the class.)

4. Hand out the research questions and initiate the group process. Please note that the in-class time required for this activity will vary based on grade level, the amount of time students can work outside class, and the amount of teacher guidance given. Expect to spend multiple days on the inquiry.

Assessment

There are several ways to analyze each data set, particularly if the research questions are modified to deal with specific populations (e.g., infants or the elderly). The examples shown in Appendix 10.1 only include one analysis per question and a few possible follow-up analyses. Students should do additional analyses to modify or strengthen their interpretations of the data.

The final report could take a number of forms, including a written lab report, a PowerPoint presentation, or a poster. The rubric shown in Student Page 10.3 can be modified for use with any of the above formats. Please fill in the "Points Possible" column before you photocopy the Student Page.

Modifications

This project is substantial enough that students may desire more public recognition than just a letter grade. Consider having students present their projects to a local health professional, parents, or other classes in the school. Alternatively, the class could send a letter to a local politician to express concern about troubling findings in their research or to recommend improvements to the real-life heat wave warning system in your area (if one exists).

Depending on the level of your students and your access to statistical software, you may want to incorporate significance testing (i.e., p-values) in the data analysis. This may help to clarify the interpretation of study results.

To greatly simplify this lesson, you could ask students to work through the figures in Appendix 10.1. Have them interpret the graphs and suggest alternate analyses. Also, ask them to think about how the heat-mortality relationships in Baton Rouge might compare with those in your geographic area.

CONCLUSION

This open-ended inquiry encourages students to take an active role in designing a study and collecting, analyzing, and interpreting data in support of that study. Furthermore, the inquiry addresses an authentic health issue and gives students the opportunity to study human populations in or near their own area. This authenticity and local relevance may be particularly helpful for engaging urban students who otherwise feel little connection to life sciences.

REFERENCES

Alexander, L. V., X. Zhang, T. C. Peterson, J. Caesar, B. Gleason, A. M. G. Klein Tank, M. Haylock, D. Collins, B. Trewin, F. Rahimzadeh, A. Tagipour, K. Rupa Kumer, J. Revadekar, G. Griffiths, L. Vincent, D. B. Stephenson, J. Burn, E. Aguilar, M. Brunet, M. Taylor, M. New, P. Zhal, M. Rusticucci, and J. L. Vazquez-Aguirre. 2006. Global observed changes in daily climate extremes of temperature and precipitation. *Journal of Geophysical Research* 111: D05109, dol.10.1029/2005JD006290.

Beniston, M., D. B. Stephenson, O. B. Christensen, C. A. T. Ferro, C. Frei, S. Goyette, K. Halsnaes, T. Holt, K. Jylhä, B. Koffi, J. Palutikof, R. Schöll, T. Semmler, and K. Woth. 2007. Future extreme events in European climate: An exploration of regional climate model projections. *Climatic Change* 81(Suppl. 1): 71–95.

Christensen, J. H., B. Hewitson, A. Busuioc, A. Chen, X. Gao, I. Held, R. Jones, R. K. Kolli, W. T. Kwon, R. Laprise, V. Magaña Rueda, L. Mearns, C. G. Menéndez, J. Räisänen, A. Rinke, A. Sarr, and P. Whetton. 2007. Regional climate projections. In *Climate change 2007: The physical science basis. Contribution of Working Group I to the Fourth Assessment Report of the Intergovernmental Panel on Climate Change*, eds. S. Solomon, D. Qin, M. Manning, Z. Chen, M. Marquis, K.B. Averyt, M. Tignor, and H. L. Miller, 847–940. Cambridge: Cambridge University Press.

Davis, R. E., P. C. Knappenberger, P. J. Michaels, and W. M. Novicoff. 2004. Seasonality of climate-human mortality relationships in US cities and impacts of climate change. *Climate Research* 26(1): 61–76.

Economic and Social Affairs, United Nations (ESA). 2005. World population prospects: The 2004 revision. Highlights. Retrieved September 18, 2007, from *www.un.org/esa/population/publications/WPP2004/2004Highlights_finalrevised.pdf*

Environmental Protection Agency (EPA). 2006. Excessive heat events guidebook. EPA 430-B-06-005. Retrieved September 18, 2007, from *www.epa.gov/heatisland/pdf/EHEguide_final.pdf*

Grimmond, S. 2007. Urbanization and global environmental change: Local effects of urban warming. *The Geographical Journal* 173(1): 83–88.

Harlan, S. L., A. J. Brazel, L. Prashad, W. L. Stefanov, and L. Larsen. 2006. Neighborhood microclimates and vulnerability to heat stress. *Social Science and Medicine* 63(11): 2847–2863.

Karl, T. R., and R. W. Knight. 1997. The 1995 Chicago heat wave: How likely is a recurrence? *Bulletin of the American Meteorological Society* 78(6): 1107–1119.

Kilbourne, E. M. 1997. Heat waves and hot environments. In *The public health consequences of disasters*, ed. E. K. Noji, 245–269. New York: Oxford University Press.

Koppe, C., S. Kovats, G. Jendritzky, and B. Menne. 2004. Heat-waves: Risks and responses. Retrieved September 18, 2007, from *www.euro.who.int/document/e82629.pdf*

Lantz, H. B. 2004. *Rubrics for assessing student achievement in science grades K–12*. Thousand Oaks, CA: Corwin Press.

Laschewski, G., and G. Jendritzky. 2002. Effects of the thermal environment on human health: An investigation of 30 years of daily mortality data from SW Germany. *Climate Research* 21(1): 91–103.

McGeehin, M. A., and M. Mirabelli. 2001. The potential impacts of climate variability and change on temperature-related morbidity and mortality in the United States. *Environmental Health Perspectives* 109 (Suppl. 2): 185–189.

National Weather Service (NWS). 2001. Winter storms: The deceptive killers. A preparedness guide. Retrieved November 27, 2007, from *www.nws.noaa.gov/om/winter/index.shtml*

National Weather Service (NWS). 2007a. 67-Year list of severe weather fatalities. Retrieved September 18, 2007, from *www.weather.gov/os/hazstats.shtm*

National Weather Service (NWS). 2007b. Storm data preparation. National Weather Service Instruction 10-1605. Retrieved November 24, 2007 from *www.nws.noaa.gov/directives/010/010.htm*

Rogot, E., P. D. Sorlie, and E. Backlund. 1992. Air-conditioning and mortality in hot weather. *American Journal of Epidemiology* 136(1): 106–116.

Trenberth, K. E., P. D. Jones, P. Ambenje, R. Bojariu, D. Easterling,

A. Klein Tank, D. Parker, F. Rahimzadeh, J. A. Renwick, M. Rusticucci, B. Soden, and P. Zhai. 2007. Observations: Surface and atmospheric climate change. In *Climate change 2007: The physical science basis. Contribution of Working Group I to the Fourth Assessment Report of the Intergovernmental Panel on Climate Change*, eds. S. Solomon, D. Qin, M. Manning, Z. Chen, M. Marquis, K.B. Averyt, M. Tignor, and H. L. Miller, 235–336. Cambridge: Cambridge University Press.

United Nations Environment Programme (UNEP). 2007. *Global Environment Outlook: Environment for development (GEO-4). Summary for Decision Makers*. Retrieved November 25, 2007, from *www.unep.org/geo/geo4/media*

OTHER RECOMMENDED RESOURCES

These additional resources were used to create the Student Pages, but are not cited in the text:

Page, R. M., G. E. Cole, and T. C. Timmreck. 1995. *Basic epidemiological methods and biostatistics: A practical guidebook*. Sudbury, MA: Jones and Bartlett Publishers.

Wassertheil-Smoller, S. 2004. *Biostatistics and epidemiology: A primer for health and biomedical professionals*, 3rd ed. New York: Springer.

Webb, P., C. Bain, and S. Pirozzo. 2005. *Essential epidemiology: An introduction for students and health professionals*. Cambridge: Cambridge University Press.

Wilkinson, P., ed. 2006. *Environmental epidemiology*. Berkshire, UK: Open University Press.

Chapter 10
Cruel, Cruel Summer

Heat waves increase from pole to poles

Student Pages

Note: Reference List for Students

For more information on references cited in the Chapter 10 Student Pages, go to teacher references on page 205.

STUDENT PAGE 10.1

Scenario Letter

[Insert agency logo and address]

[Insert today's date]

[Insert address of your school]

Dear Scientists:

Climate scientists predict that in the near future, heat waves will increase in frequency, duration, and severity. We are concerned about the potential impacts of climate change on the effectiveness of our heat wave warning system. We need your assistance in collecting basic information on the relationship between climate change and the risk of heat-related deaths. We also would appreciate recommendations on key elements to include in our warning system and associated social support services.

After you finish Epidemiology 101, our representative will provide you with a list of potential research questions and our expectations. These questions represent gaps in our current knowledge. If, in your expert opinion, there are other topics that need to be addressed, please obtain approval from our representative.

Thank you for your efforts to protect our citizens. We look forward to hearing from you by *[insert due date of assignment]*.

Sincerely,

[Insert fictional name and signature
of an agency representative]

STUDENT PAGE 10.2

Epidemiology 101

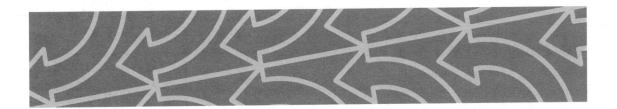

Welcome to Epidemiology 101, a crash course in the study of the distribution and causes of health problems. By the end of this course, you should be able to

- define the words *epidemiologist* and *risk*,
- calculate mortality rate,
- calculate relative risk, and
- give examples of how epidemiologists study the effects of climate change on health.

What Is an Epidemiologist?

Epidemiologists are disease detectives. In other words, they try to answer the following questions:

- Who is becoming ill?
- Where is the illness occurring?
- What is the root cause of the illness?
- How can we reduce the risk of becoming ill in the future?

Epidemiologists can study people, other animals, or even plants. Today, we are concerned with the human branch of epidemiology.

Research Techniques

In a given year, the number of human deaths is affected by hundreds of factors, from the state of the economy to natural disasters. Because human society is so complex, epidemiologists use data from studies on the short-term effects of weather to predict the long-term effects of climate.

One of the basic calculations used in epidemiology is **mortality rate**, or the number of deaths in a given population. Because the mortality rate for most illnesses and injuries is, thankfully, very low, epidemiologists don't just divide the number of deaths by the number of people. Instead they multiply the proportion by a large factor, such as 1,000 or 10,000. It's easier to understand a proportion like 0.001 when you think of it as 1 death per 1,000 people or 10 deaths per 10,000 people.

$$\text{Mortality rate} = \left(\frac{\text{\# Deaths}}{\text{\# People in study population}} \right) \times 1{,}000$$

(If you wanted to calculate the rate per 10,000 people, you would substitute 10,000 for 1,000).

There are two ways to predict risk, or how many people will die in future heat waves. The first way is to calculate **relative risk**. A **risk factor** is a characteristic of a certain population that makes that population susceptible to disease. There are a number of risk factors for heat-related death, including age, poverty, other illnesses, and geographic location. Relative risk

$$\text{Relative risk} = \frac{\text{Mortality rate in population with risk factor}}{\text{Mortality rate in population with no risk factor}}$$

is a measure of how much more likely it is that an individual will become ill if he or she has a certain risk factor. Here is the equation:

For instance, if the mortality rate among elderly people (old age being a risk factor) during a 10-day heat wave is 8 per 10,000, and the mortality rate among high school students is 4 per 10,000, the relative risk for elderly people would equal 2. This means if a 10-day heat wave strikes the same geographic area in the future, elderly people are 2 times more likely to die from heat than are high school students.

SAMPLE PROBLEM 1

Residents of Springfield (population: 30,720) have more access to air conditioning than residents of the neighboring town of Shelbyville (population: 25,350). In the summer of 2007, the weather was unusually hot and humid for one week. During that week, 15 people died in Springfield and 40 died in Shelbyville.

ANSWER THESE QUESTIONS:

1. What was the mortality rate during the heat wave in Springfield?

2. What was the mortality rate during the heat wave in Shelbyville?

3. What is the relative risk of living in Shelbyville during a future heat wave?

The second way to predict the future risk of deaths during heat waves is with **regression**. In regression, the relationship between two variables is summarized with a **line of best fit** (Figure 10.6). The line is described by an equation ($y = mx + b$) and a number called the **coefficient of determination**. This coefficient (abbreviated as R^2) is a measure of how strong the relationship is between the variables or, in other words, how well values of the x variable predict values of the y variable. An R^2 of 1 means the line of best fit is a perfect description of how the x and y variables are related. An R^2 of 0 means there is no relationship between x and y. You can calculate the line of best fit and R^2 with software such as Excel. Epidemiologists commonly perform regressions of mortality against time, temperature (Figure 10.6), or age.

Figure 10.6

Sample of mortality analysis by temperature.

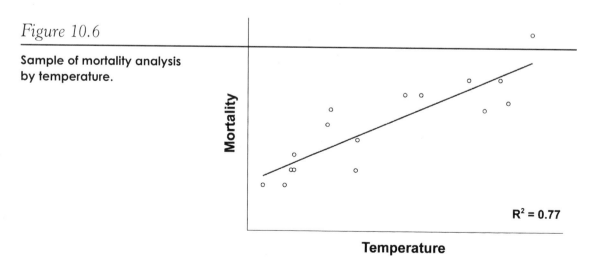

$R^2 = 0.77$

Temperature

SAMPLE PROBLEM 2

In Springfield, the relationship between mortality rate (per 10,000) and temperatures between 30°C and 40°C is described by the following equation: $y = 0.58x - 12.9$. The R^2 is 0.56.

ANSWER THESE QUESTIONS:

1. What is the mortality rate when the temperature is 30.8°C?

2. What is the mortality rate when the temperature is 38.7°C?

3. How well does temperature predict mortality? In other words, is the relationship between the two variables weak, moderate, or strong?

Finally, there are a number of ways to display epidemiological data. In addition to scatter-plots, like Figure 10.6, you could use tables, bar graphs, or maps (see Figure 10.7 for examples).

Figure 10.7

Examples of ways to display mortality data.

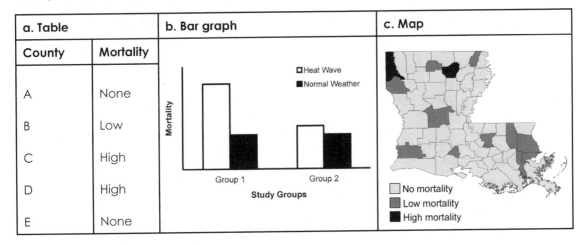

STUDENT PAGE 10.3

Research Questions and Expectations

Possible Questions

- How are mortality rates in our area related to temperature?
- Is heat-related mortality risk higher in early summer or late summer?
- Does the homicide rate increase during hot years?
- Is there an increase in the risk of death due to infectious disease during heat waves in our area? (You may want to consider only one disease.)

Expectations

You might be tempted to answer your research question with just one analysis. Recall from Epidemiology 101, however, that human death rates are complex! How can you be sure an association you have found reflects a cause-and-effect relationship? Use the checklist in Figure 10.8 to guide you as you analyze your data and make interpretations.

Figure 10.8

Data analysis checklist.

The data show a strong relationship between the variables (i.e., a large R^2).	
There is a logical biological reason for the variables to have a strong relationship.	
The relationship between the variables is consistent with work by other researchers. (Look for studies similar to or the same as yours. You also may discuss your interpretations with other groups of students.)	
Other explanations for the relationship between the variables are not as logical or can be ruled out.	

Your project will be graded with the form below (Figure 10.9). When you have completed your project, fill in the points for the "Self" column and return it, the checklist, and your project to your teacher.

Figure 10.9

Grading form.

Names of students in group:				
Performance criteria	**Points possible**	**Self**	**Teacher**	**Comments**
The study area and population are clearly defined.				
A prediction developed from information presented in class and background readings is clearly stated.				
A logical, step-by-step set of procedures used for conducting the study is listed.				
Appropriate statistical procedures were selected and applied.				
Appropriate graphic display techniques were used to display the data.				
Appropriate inferences were made based on the data interpretation. Additional analyses were conducted, if necessary, to clarify uncertain inferences.				
A logical recommendation that is based on the data analysis and background reading is provided to the health agency.				

Source: Performance criteria modified from Lantz 2004.

APPENDIX 10.1

Examples of Data Analysis for "Cruel, Cruel Summer"

Question 1: How Are Mortality Rates in Our Area Related to Temperature?

STUDY POPULATION

Residents of the city of Baton Rouge, Louisiana.

PREDICTION

Mortality rates will increase as maximum temperature increases.

METHODS

- Regression analysis.
- Graphed maximum temperature (TMAX) against observed mortality rate for each week of June-August from 2001 to 2005.

RESULTS

Our prediction was not supported (Figure 10.10). In June, July, and August of 2001–2005, mortality rates decreased as maximum weekly temperatures increased. (Note, however, that the R^2 is very small.)

Figure 10.10

Relationship between maximum summer temperature and weekly mortality rate of residents of Baton Rouge, Louisiana, 2001–2005.

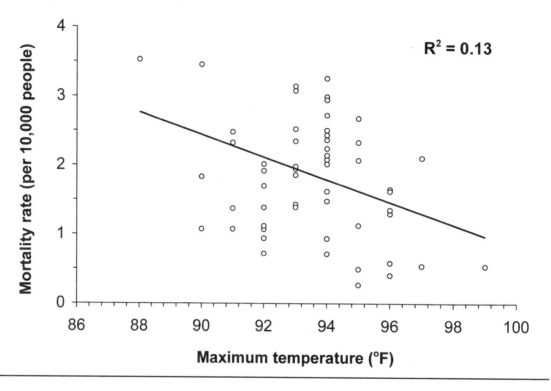

INTERPRETATION

Nearly every household and business in Baton Rouge has air conditioning, and people may spend more time indoors as temperatures rise. In other words, the use of air conditioning may be preventing the expected relationship between maximum temperature and mortality. Alternatively, people may be more acclimated to—and therefore less stressed by—warm weather in late summer, when temperatures peak.

OTHER POSSIBLE ANALYSES

- Examine seasonal trend in mortality rates. In other words, does the relationship between maximum temperature and mortality differ between spring and summer? Summer and autumn?
- Expand range of temperatures to include cooler weather.
- Repeat analysis using both heat and humidity.
- Collect data to investigate whether people in Baton Rouge spend more time indoors when it's hot out.

218

Question 2: Is Heat-Related Mortality Risk Higher in Early Summer or Late Summer?

STUDY POPULATION

Residents of the city of Baton Rouge, Louisiana.

PREDICTION

There will be more heat wave deaths early in the summer, because people are not yet acclimated to hot weather.

METHODS

- Calculated relative risk.
- Looked for years with heat waves at the beginning of June or July and the end of August; could only get data as far back as 1998.

RESULTS

Our prediction was not supported (Figure 10.11). In 2 of 3 years, mortality was higher during late summer heat waves than during early summer heat waves.

Figure 10.11

Mortality rate and relative risk of death during early summer vs. late summer heat waves in Baton Rouge, Louisiana, summers of 1998, 2000, and 2005.

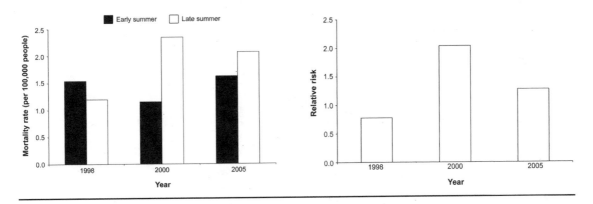

INTERPRETATION

The late summer heat waves in 1998 and 2005 may have been longer or more severe (hotter and/or more humid) than the early summer heat waves.

OTHER POSSIBLE ANALYSES

- Compare severity and duration of early heat waves to late heat waves.
- Repeat analysis using both temperature and humidity.

Question 3: Does the Homicide Rate in Our Area Increase During Hot Years?

STUDY POPULATION

Residents of East Baton Rouge Parish, Louisiana.

PREDICTION

The risk of homicide will increase during hot years.

METHODS

- Regression analysis.
- Calculated mean TMAX from June–August for each of 1999–2004.

RESULTS

Our prediction was weakly supported (Figure 10.12). The homicide risk was greater in years with higher summer temperatures.

Figure 10.12

Relationship between homicide rate and summer temperatures, East Baton Rouge Parish, Louisiana, 1999–2004.

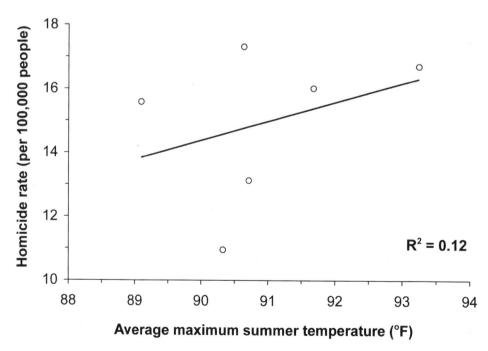

INTERPRETATION

People become more aggressive when they are stressed by heat, possibly because of hormonal changes (e.g., increased testosterone). However, given the low R^2, it appears factors other than heat are primarily responsible for the homicide rate.

OTHER POSSIBLE ANALYSES

- See if this relationship holds for other parishes (counties) in Louisiana or other Gulf Coast states.
- Look for economic trends in each year.
- Survey people about changes in their emotional state during extremely hot weather.

Climate Change From Pole to Pole

Question 4: Is There an Increase in Mortality Risk Due to Infectious Disease During Heat Waves?

STUDY POPULATION

Residents of the city of Baton Rouge, Louisiana.

PREDICTION

Risk of death from pneumonia and influenza will increase during heat waves.

METHOD

- Calculated relative risk.
- Compared mortality rates from five weeks in August and September in each of 1998–2005. (Four of these years were abnormally hot, and the other four were normal.)

RESULTS

Our prediction was supported (Table 10.1). In Baton Rouge, the risk of death from pneumonia and influenza is about two times higher during heat waves than during normal weather.

INTERPRETATION

The flu virus might spread more easily during hot weather. However, it is more likely that individuals who died of pneumonia or the flu already were stressed by the hot weather. Regardless, this is not a significant public health threat, as fewer than 2 people in 100,000 are dying from these diseases in the summer.

OTHER POSSIBLE ANALYSES

- Do a regression between pneumonia/ flu mortality and temperature. Should probably include cool months, since flu is more common in the winter than in the summer.
- Repeat this analysis with only age groups most sensitive to flu (i.e., elderly and very young) or by economic status (i.e., people who probably don't have health insurance).
- Repeat analysis using temperature *and* humidity.

Table 10.1

Deaths from influenza and pneumonia in August and September, 1998–2005, Baton Rouge, Louisiana.

WEATHER	MORTALITY RATE (PER 100,000)	RELATIVE RISK OF MORTALITY
Normal	0.7	--
Extreme heat	1.4	2.1

Glossary

A

acclimatization. Physiological adjustment to an environmental stress such as extreme heat or drought.

aerosol. Fine particle of liquid or solid that is suspended in the atmosphere.

albedo. The proportion of solar radiation reflected by an object.

allergen. A normally harmless substance such as food or pollen that is recognized as an invader by the immune system. Can cause skin rashes, excess mucus production, difficulty breathing, and other symptoms. *See also* **allergic rhinitis.**

allergic rhinitis. *Also called* hay fever. Inflammation of mucus membranes of the nose and eyes in response to **allergens** such as pollen, mold, or dust mites. Can occur seasonally or year-round.

anthropogenic. Caused by humans. Often pertains to the effect of humans on the environment.

B

biome. Large areas of the planet defined by unique functional (i.e., morphological or physiological) groups of plants and animals adapted to a particular regional **climate**. Examples include **tundra,** grassland, **tropical rainforest,** and desert.

bleaching. *See* **coral bleaching.**

C

carbon isotope ratio. The ratio of two naturally occurring isotopes of carbon (^{12}C and ^{13}C). Can be used to identify the source (i.e., fossil fuels, volcanoes, etc.) of carbon dioxide in the atmosphere.

carbon sink. Area of the Earth that temporarily removes carbon from the atmosphere through geochemical and biological processes.

carrion. The rotting carcass of an animal. Is an important food source for scavengers and many carnivores.

climate. The state of the atmosphere over years to decades. Describes average values of temperature, precipitation, cloudiness, and other meteorological elements, as well as their variability, seasonality, and extremes. *See also* **macroclimate** and **microclimate.**

climate change. A significant and persistent change in any climate element (e.g., precipitation or clouds), at any spatial scale (e.g., regional, continental, or global). In the past century, has been caused by human activities and nonanthropogenic processes such as volcanic eruptions. *See also* **global warming.**

climate deviation. *Also called* climate anomaly. The difference between climate measurements at a given time (e.g., June 1987) and weather station (e.g., Boulder, Colorado) and the **climate normal** for the same station.

climate normal. A long-term mean in a climate element such as temperature at a given location. Is calculated over a 30-year period.

climate sensitivity. An estimate of the mean increase in global temperature that will result if the preindustrial concentration of CO_2 is doubled. The current best estimate, according to the Intergovernmental Panel on Climate Change, is 3°C.

coefficient of determination (R^2). An index of how well values of one variable predict values of a second variable. An R^2 of 1 indicates a perfect straight-line relationship between two variables. An R^2 of 0 means there is no relationship. *See also* **regression analysis.**

community. A group of interacting **populations** living in a given area.

condensation nuclei. **Aerosols** that form the core of cloud droplets. With the addition of sufficient water molecules, cloud droplets can fall from clouds as rain or snow.

conduction. The transport of energy by random molecular motions (i.e., no mass is exchanged between high-energy and low-energy matter). Is most effective in solids. *See also* **convection.**

convection. The transport of heat and/or moisture by the movement of a fluid such as air (i.e., mass is exchanged between high-energy and low-energy matter). *See also* **conduction.**

coral bleaching. The expulsion of symbiotic algae from reef-building corals. Occurs when corals are under stressful conditions such as high temperature or low light. Bleached corals appear white or pale in color.

Coriolis force. The deflection of air and ocean currents to the right in the Northern Hemisphere and to the left in the Southern Hemisphere. Caused by Earth's rotation.

correlation analysis. A statistical technique used to measure the strength of a relationship between two variables. A positive correlation occurs if increasing values of one variable are associated with increasing values of the other variable. A negative correlation occurs if increasing values of one variable are associated with decreasing values of the other variable. *See also* **regression analysis.**

cryosphere. The frozen portions of Earth's surface, including sea ice, snow, glaciers, and **permafrost.**

E

easterlies. *Also called* trade winds. Winds that blow from the northeast or southeast, starting at the **subtropical highs** (high atmospheric pressure) and ending at the **equatorial trough** (low atmospheric pressure).

ecosystem. A **community** of organisms plus their associated physical environment.

ectotherm. An animal, such as a frog or an insect, that cannot control its body temperature internally (i.e., its body temperature is primarily dictated by environmental conditions).

El-Niño Southern Oscillation (ENSO). An irregular cycle of sea surface temperatures and atmospheric pressure in the western and central Pacific that oscillates between a warm phase (El Niño) and a cold phase (La Niña). ENSO can affect temperature and precipitation patterns worldwide.

environmental gradient. A gradual transition from one environmental state (e.g., hot and dry) to another (e.g., cold and wet). Can be used as a "natural experiment" to investigate how organisms are affected by changes in one or more abiotic factors.

epidemiology. The study of the distribution, frequency, and causes of illness, injury, or disease in human populations.

equatorial trough. A belt of low atmospheric pressure near the equator where the **easterlies** from the Northern and Southern Hemispheres meet.

evapotranspiration. The movement of water from the soil (through evaporation) and from living plants (through transpiration) to the atmosphere.

extreme heat day. A day on which the maximum air temperature exceeds a local threshold. Common thresholds include the 85th, 90th, and 95th percentile.

F

faculae. Areas on the Sun's surface that are hotter than average and therefore emit more radiation. Are generally associated with **sunspots.**

feedback. Internal cycle in the climate system that either amplifies or diminishes the effects of climate warming. *See also* **forcing agent.**

fledgling. A young bird that can fly but is still under the care of its parents.

forcing agent. A substance or process that affects Earth's energy balance by altering the amount of incoming solar radiation, reflected solar radiation, or outgoing longwave radiation. *See also* **radiative forcing.**

G

gigaton (Gt). A metric unit of mass equal to 1 trillion kilograms.

global warming. An increase in mean global temperatures that can contribute to changes in other climate elements, such as precipitation or the severity of storms. *See also* **climate change.**

greenhouse effect. A process by which radiation is "recycled" between Earth and the atmosphere because of the selective absorption of radiation by greenhouse gases. Without the natural greenhouse effect, Earth's temperature would be about 33°C cooler than it is now. The greenhouse effect has been enhanced by the release of greenhouse gases through fossil fuel use, land use changes, and burning of vegetation. *See also* **greenhouse gas**.

greenhouse gas. A gas in Earth's atmosphere that slows the loss of longwave radiation to space. Greenhouse gases transmit short-wave radiation (i.e., allow it to pass through unchanged) but absorb some wavelengths of longwave radiation. Carbon dioxide, methane, and other gases that stay in the atmosphere for years to decades are called long-lived greenhouse gases. Water vapor is a short-lived greenhouse gas. *See also* **greenhouse effect**.

H

heat index. *Also called* apparent temperature. A measure of how hot the weather "feels" based on heat and humidity combined.

heat wave. A period of extremely hot weather that also may be accompanied by high humidity. There is no universal agreement on how long the "period" is or what constitutes "extremely hot."

historical sources. Human-made records of climate, including agricultural records and the diaries of mariners. Can be used to reconstruct past climates and to investigate **climate change.**

I

instrumental record. Direct measurements of **climate** elements such as temperature, precipitation, humidity, wind speed and direction, and atmospheric chemistry. Also includes indirect indicators of **climate change,** such as sea level. Provides information about the present and the recent past (i.e., approximately 150 years of Earth's past).

Inuit. Aboriginal people of the Canadian Arctic. In Canada, the term "Eskimo" is considered derogatory and is no longer used.

K

keystone species. A species that affects its **community** or **ecosystem** in a much larger way than expected based on abundance alone. Keystone species are usually top predators.

L

La Niña. *See* **El-Niño Southern Oscillation.**

latent heat transfer. The loss or gain of heat energy that occurs when water changes phase between a liquid, gas, or solid. When water evaporates, it gains energy from the surface it is evaporating from, and that surface cools. When water condenses in the atmosphere, it loses the energy it gained during evaporation and heats the surrounding air.

limiting factor. A component of an organism's

environment that constrains that organism's life processes. Can be abiotic (e.g., temperature) or biotic (e.g., competition with another organism).

line of best fit. A line that summarizes the statistical relationship between two variables. May be straight or curved, depending on the analysis used. Commonly used to display results of **regression analyses**. *See also* **trend line.**

long-lived greenhouse gas. *See* **greenhouse gas.**

longwave radiation. *Also called* far infrared radiation. Electromagnetic waves 4–100 microns long that are emitted by the atmosphere and Earth's surface. *See also* **radiation** and **shortwave radiation.**

M

macroclimate. The prevailing **climate** of a region. Can extend for hundreds of kilometers. *See also* **microclimate.**

microclimate. The prevailing **climate** of a small area. Can range in size from a bubble of air surrounding an organism to an entire hillside. *See also* **macroclimate.**

Milankovitch cycles. Changes in Earth's orbit that appear to trigger the initiation and temination of ice ages.

model. Numerical representation of a complex system. Used to define important relationships in a system and to make realistic projections about future changes to the system. For example, climatologists can use models to forecast the effect of increasing carbon dioxide emissions. Ecologists can use models to understand how organisms respond to increasing temperature or to forecast the future **range** of a species under a warmer **climate.**

mortality rate. The number of deaths in a population of people exposed to a specific health risk. Usually expressed as the number of people per 1,000; 10,000; or 100,000 people.

N

non-native species. *Also called* invasive, exotic, or alien species. Any organism that is transported accidentally or intentionally by humans to a new habitat.

novel ecosystem. A combination of species that has not previously occurred within a given biome.

P

paleoclimatology. The study of ancient **climates.**

paleoecology. The study of how prehistoric species interacted with one another and their environment.

parts per million (ppm). A measure of concentration calculated as a unit of one substance dissolved in a million units of a second substance.

permafrost. A layer of subsurface soil that remains frozen year-round. Typically found at high latitudes and high altitudes. Is part of Earth's **cryosphere.**

phenology. The timing of life cycle events such as flowering or seasonal migration. Can be controlled by physical factors, such as temperature, moisture, or day length, and by biological factors, such as hormones.

population. A group of individuals of one species living in a given area at a given time.

problem bears. Bears that have become accustomed to people and so are not afraid of them. They can threaten human life and property.

proxy. A physical, biological, or chemical entity that can be used to infer the state of past **climates** or to assess recent changes in climate.

Examples include tree rings, fossils, pollen, and modern communities of plants or animals.

R

radiation. Electromagnetic energy that travels in the forms of waves (also called rays). *See also* **longwave radiation** and **shortwave radiation.**

radiative forcing. An index of how much a **forcing agent** has changed Earth's energy balance. A positive value for a forcing agent means the agent has resulted in greater energy gains than losses (i.e., warming effect). A negative value for a forcing agent means the agent has resulted in greater energy losses than gains (i.e., cooling effect).

range. The geographical area in which a **population** or species lives. In field guides, species ranges are shown as continuous areas with static boundaries. In reality, ranges typically are discontinuous and dynamic (i.e., the edges are always changing).

recruitment. Addition of potential breeders to a population through births and subsequent maturity.

regression analysis. A statistical technique used to describe or predict how changes in one variable (x) result in changes in a second variable (y). Unlike **correlation analysis**, the x variable in regression usually has been experimentally manipulated (rather than passively observed/measured). *See also* **coefficient of determination** and **line of best fit.**

relative risk. A measure of how much more likely it is that an individual with a certain **risk factor** will become ill, compared with someone without it.

risk factor. A characteristic that makes a person or group of people susceptible to disease.

S

sensible heat transfer. The transfer of energy by **conduction** and **convection** from a surface to the air. Sensible heat increases speed at which air molecules move and thus raises the air temperature.

scientific consensus. An agreement among scientists that a scientific explanation is logical, realistic, and well-supported by multiple lines of evidence.

short-lived greenhouse gas. *See* **greenhouse gas.**

shortwave radiation. *Also called* visible light. Electromagnetic waves 0.4–0.7 microns long that are emitted by the Sun. *See also* **radiation** and **longwave radiation.**

specific humidity. A measure of how much water vapor is in a given amount of atmosphere. Usually expressed as a ratio of the mass of water vapor to the total mass of water vapor plus air.

stabilizing selection. Selection against extreme phenotypes. Under this type of natural selection, evolutionary changes in populations may be impeded.

structural uncertainty. *See* **uncertainty.**

subtropical highs. Belts of high atmospheric pressure at about 30°N and 30°S. Warm air moving poleward from the equator loses heat and moisture and settles near the surface in these belts.

sunspot. A relatively cool, dark area on the surface of the Sun that is created by magnetic disturbances and that can last days to weeks. Sunspots are generally associated with **faculae.** When the number of sunspots increases, the Sun emits a greater amount of shortwave radiation.

T

temperature dependent sex determination. A mode of sex determination in which the incubation temperature of eggs, rather than genetic makeup, determines the gender of an organism. Occurs in many species of reptiles, including most turtles.

traditional ecological knowledge (TEK). Information and understanding about the environment that has been acquired by people living in close association with the natural world.

trend line. A line that summarizes the change in a variable over time. May be straight, curved, or irregularly shaped depending on the statistical analysis used. *See also* **line of best fit.**

trophic mismatch. A lack of correspondence between the life cycle of an organism and the life cycle of that organism's food source.

tropical rainforest. A **biome** found at or near the equator that experiences high heat and humidity. The vegetation in this biome (primarily broad-leaved trees) is highly diverse, productive, and structurally complex.

tundra. A highly seasonal **biome** found around the Arctic Circle, on sub-Antarctic islands, and at high altitudes. The vegetation in this biome is typically low-lying and tolerant of extended periods of drought and cold (e.g., lichens and mosses).

U

uncertainty. A lack of certainty about the accuracy or completeness of a scientific result or explanation. Can occur during the study of **climate change** when data are inaccurate or incomplete (i.e., value uncertainties) or when explanations or models are missing relevant processes or relationships (i.e., structural uncertainties).

V

value uncertainty. *See* **uncertainty.**

W

westerlies. Winds that blow from the southwest or northwest, starting at the **subtropical highs** and ending near the poles.

Index

234

National Science Teachers Association